The Geography and Remote Sensing Analysis of Sri Lanka

Xianhu Wei

The Geography and Remote Sensing Analysis of Sri Lanka

Hubei Science & Technology Press

Springer

Xianhu Wei
Aerospace Information Research Institute
Chinese Academy of Sciences
Beijing, China

ISBN 978-981-33-4626-0 ISBN 978-981-33-4627-7 (eBook)
https://doi.org/10.1007/978-981-33-4627-7

Jointly published with Hubei Science and Technology Press Co., Ltd.
The print edition is not for sale in China (Mainland). Customers from China (Mainland) please order the print book from: Hubei Science and Technology Press Co., Ltd.

This Springer imprint is published by the registered company Springer Nature Singapore Pte Ltd.
The registered company address is: 152 Beach Road, #21-01/04 Gateway East, Singapore 189721, Singapore

Preamble

With remote sensing images as the main data source and reference to socio-economic statistics and field survey, the book has carried out the status survey, dynamic monitoring and comprehensive analysis on Sri Lanka's land, vegetation, surface water, ocean and other environmental resources as well as economic, transportation, urban, agricultural and tourism development. The book can provide scientific evidences for an objective, accurate, systematic and comprehensive understanding of the status quo and progress of Sri Lanka's ecological environment and socio-economic development, and for further policies for the protection of the ecological environment and biodiversity. Moreover, it offers information for the comprehensive cooperation in science and technology between two countries.

Introduction

Located at the intersection of the east-west passage on the Indian Ocean, Sri Lanka is the converging point between the south branch line of the ancient Silk Road and the Maritime Silk Road and is one of the important countries for China's Belt and Road initiative.

To respond to the calls of the Belt and Road initiative, Chinese Academy of Sciences (CAS) has launched a series of pioneering work combined with its own characteristics of scientific and technological innovation, including such joint projects in Sri Lanka as China-Sri Lanka Joint Center for Education and Research and China-Sri Lanka Joint Research and Demonstration Center for Water Technology (hereinafter referred to as China-Sri Water Center). Directed by the Research Center for Eco-Environmental Sciences, CAS, China-Sri Water Center is an international platform for education and research jointly constructed by State Ministry of City Planning and Water Supply, National Water Supply and Drainage Board and University of Peradeniya. The Chinese contributors to the platform include Institute of Remote Sensing and Digital Earth and other six research institutes of CAS, and Peking University First Hospital, Beijing Center for Diseases Prevention and Control and Peking University. Major life issues in Sri Lanka's drought areas are expected to be addressed, such as Chronic Kidney Disease of unknown etiology (CKDu) and drinking water safety. The platform is designed to bring greater benefits for Sri Lanka's people and make more contributions to promoting our friendship through joint research, talent training and technical demonstration.

Under the direction of "China-Sri Lanka Water Center", the joint research group of water resources has carried out remote sensing monitoring research on the soil, water and vegetation in Sri Lanka based on spatial information technology, which has laid a foundation for the research on the rational use and management of water resources and the guarantee of drinking water safety. The book, based on the data collection and collation from relevant remote sensing monitoring in Sri Lanka, is divided into three parts. The first section outlines the geography of Sri Lanka by remote sensing images, traditional statistical, GIS spatial analysis and mapping. Mainly with remote sensing images, the second section gives an introduction and analysis of the current situation and the spatio-temporal variation characteristics of natural resources and ecological environment such as land use, vegetation cover, rivers and basins, lakes and reservoirs and marine. The final part, mainly based on remote sensing images and statistics, demonstrates and analyzes the process and the law of its economic development and urban expansion. Also through the typical remote sensing images and GIS spatial analysis and mapping, the final section directly shows the spatial distribution characteristics of its transportation and crops, and the unique landscape of world heritage sites. The book provides data for further research on resource utilization and environmental protection in Sri Lanka, and information for the construction of China-Sri Lanka Water Center, the macro decision-making of the governments and the comprehensive cooperation in technology and ecological environment along the Belt and Road scope.

AIR Aerospace Information Research Institute, CAS

Research Center for Eco-Environmental Sciences, CAS

Contents

National Geography

1 Location

Sri Lanka, an island country on the Indian Ocean between 5° 55′ N–9° 50′ N and 79° 42′ E–81° 53′ E, is located at the southern tip of the South Asian Subcontinent. It is separated from the Indian peninsula by the Palk Strait and is close to the equator in the south. It has an area of 65,600 km^2, with 435 km at length and 225 km at width. The pear-shaped country is known as "the Pearl of the Indian Ocean" (Fig. 1).

2 Administrative Divisions

The country has 9 provinces and 25 districts. The provinces are Western, Central, Southern, North Western, Northern, North Central, Eastern, Uva and Sabaragamuwa. Under the administration of the District Secretary appointed by the central government, the districts are divided into 331 Divisional Secretary's Divisions, which are further separated into Grama Niladhari Divisions, the smallest administrative level. There are 14,021 GN Divisions across the country.

Colombo, the capital of Sri Lanka, is its political, commercial and cultural center (Fig. 2 and Table 1).

X. Wei, *The Geography and Remote Sensing Analysis of Sri Lanka*,
https://doi.org/10.1007/978-981-33-4627-7_1

Fig. 1 Sri Lanka's remote sensing image

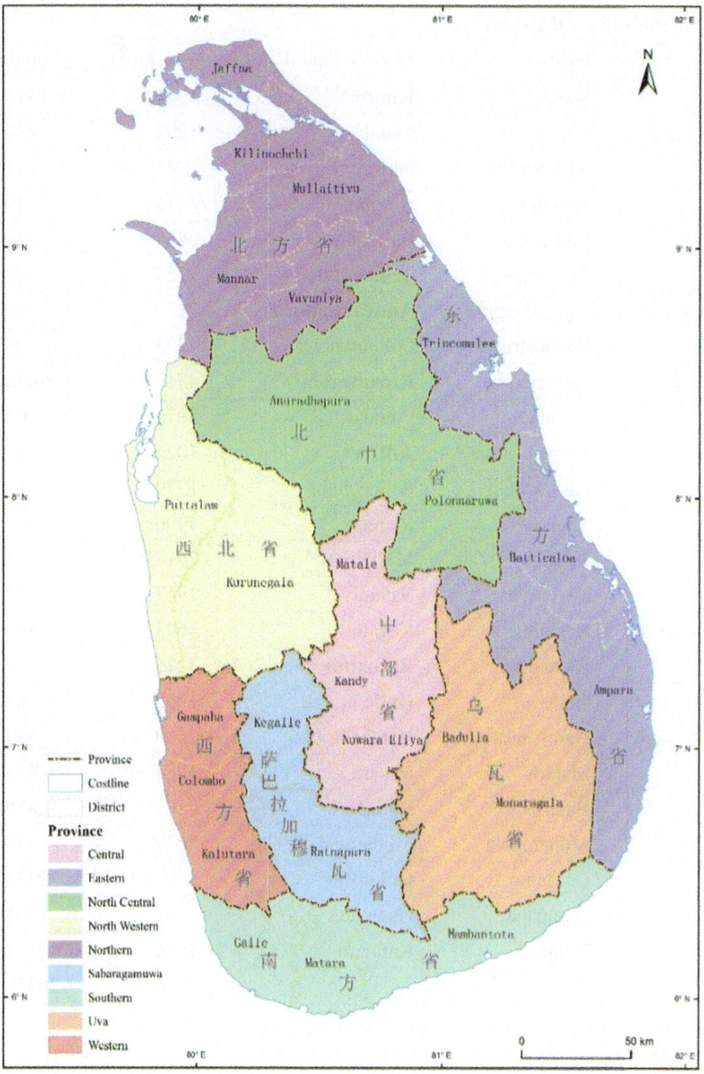

Fig. 2 Administrative divisions. *Source* Database of Global Administrative Areas (GADM), http://gadm.org

Table 1 Administrative divisions

Province	District	District capital	Area (km^2)	Population
Central	Kandy	**Kandy***	1940	1375382
	Matale	Matale	1993	484531
	Nuwara Eliya	Nuwara Eliya	1741	711644
Eastern	Ampara	Ampara	4415	649402
	Batticaloa	Batticaloa	2854	526567
	Trincomalee	**Trincomalee***	2727	379541
North Central	Anuradhapura	**Anuradhapura***	7179	860575
	Polonnaruwa	Polonnaruwa	3293	406088
North Western	Kurunegala	**Kurunegala***	4816	1618465
	Puttalam	Puttalam	3072	762396
Northern	Jaffna	**Jaffna***	1025	583882
	Kilinochchi	Kilinochchi	1279	113510
	Mannar	Mannar	1996	99570
	Mullaitivu	Mullaitivu	2617	92238
	Vavuniya	Vavuniya	1967	172115
Sabaragamuwa	Kegalle	Kegalle	1693	840648
	Ratnapura	**Ratnapura***	3275	1088007
Southern	Galle	**Galle***	1652	1063334
	Hambantota	Hambantota	2609	599903
	Matara	Matara	1283	814048
Uva	Badulla	**Badulla***	2861	815405
	Monaragala	Monaragala	5639	451058
Western	Colombo	**Colombo***	699	2324349
	Gampaha	Gampaha	1387	2304833
	Kalutara	Kalutara	1598	1221948
Total			**65610**	**20359439**

*The city marked with an asterisk is the provincial capital city
Source Census Data in 2019 from the Department of Census and Statistics, Sri Lanka

3 Climate

Sri Lanka has a tropical monsoon climate with the southwest and northeast monsoon. It's like summer all the year, with the only southwest monsoon pluvial period from May to August and northeast monsoon drought period from November to February.

3.1 Rainfall

The spatial-temporal distribution of rainfall in Sri Lanka is mainly affected by monsoon and convective precipitation with a spatially varying distribution. The yearly rainfall gradually decreases from southwest to north and east, with the highest in southwest region and the lowest in southeast and northwest coastal areas. Among them, the southwest has abundant rainfall, with the yearly average annual of 2500–5000 mm; the west slope zone in the Central Highlands of Sri Lanka has the highest annual rainfall of more than 5000 mm while the northwest and southeast have relatively less of 1250 mm and the driest regions have an average annual rainfall of less than 900 mm.

3.2 Temperature

The average annual temperature is relatively high in most parts of the country and the effects of altitude on temperature differs by region. For example, in the central mountainous area temperatures are significantly lower than that in the surrounding low altitude area. The average temperature decreases from 27.5 °C in coastal lowlands to 16 °C in the Central Highlands, 1900 m above sea level. In coastal areas, the average maximum temperature is 31.3 °C and the minimum 23.8 °C; In mountain area, 26.1 °C and 16.5 °C respectively. The monthly average temperature is slightly different, which mainly depends on the seasonal variation of the sun position and rainfall intensity. Generally, the monthly average temperature is the lowest in January and the highest in April and August (Figs. 3 and 4).

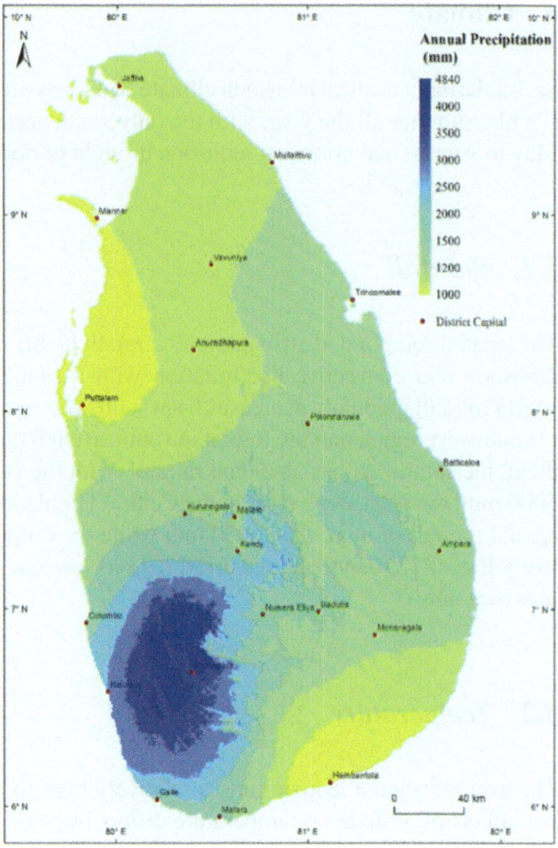

Fig. 3 Distribution of annual rainfall from 1970 to 2000. *Source* Department of Meteorology, Sri Lanka

3.3 Climate Seasons

Influenced by the southwest and northeast monsoon as well as topography, the climate in Sri Lanka can be divided into four seasons.

IM1, First Inter-monsoon Season from March to April

The rainfall is less: 100–250 mm in most areas; below 100 mm in Jaffna Peninsula; more than 300 mm in the southwest of the Central Highlands, and more than 750 mm in some parts.

SWM, Southwest-Monsoon Season from May to September

Rainfall increases, ranging from 100 to 3000 mm. It is between 1000 and 1600 mm in the southwest, over 3000 mm in the southwest slope zone of the Central Highlands and below 200 mm in the northwest and southeast.

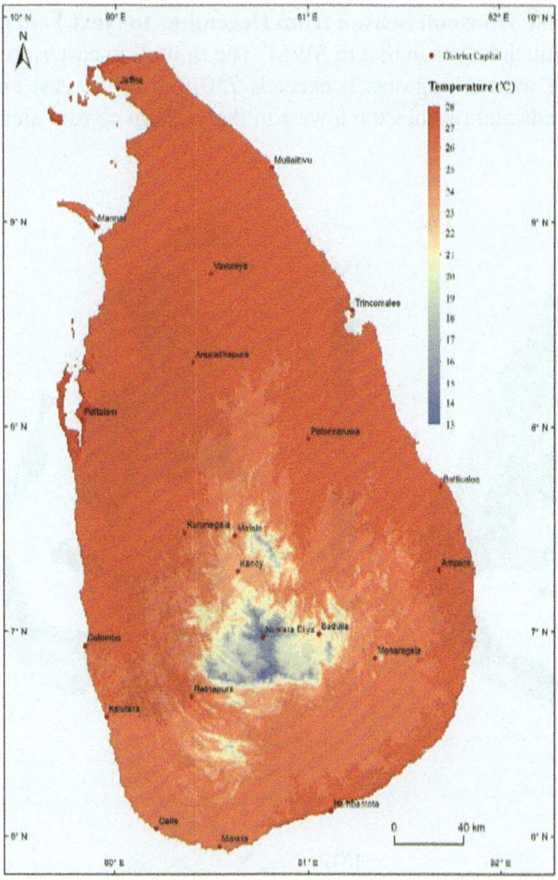

Fig. 4 Distribution of average annual temperature

IM2, Second Inter-monsoon Season in October and November

With the most balanced distribution, the rainfall of almost the whole island exceeds 400 mm with southwest slope ranging from 750 to 1200 mm.

NEM, Northeast Monsoon Season from December to Next February

The rainfall is much less than that in SWM. The rainfall in eastern regions is higher than that in the western regions. It exceeds 750 mm in the east and north of the Central Highlands and reaches the lowest in the western coastal area near Puttalam (Fig. 5).

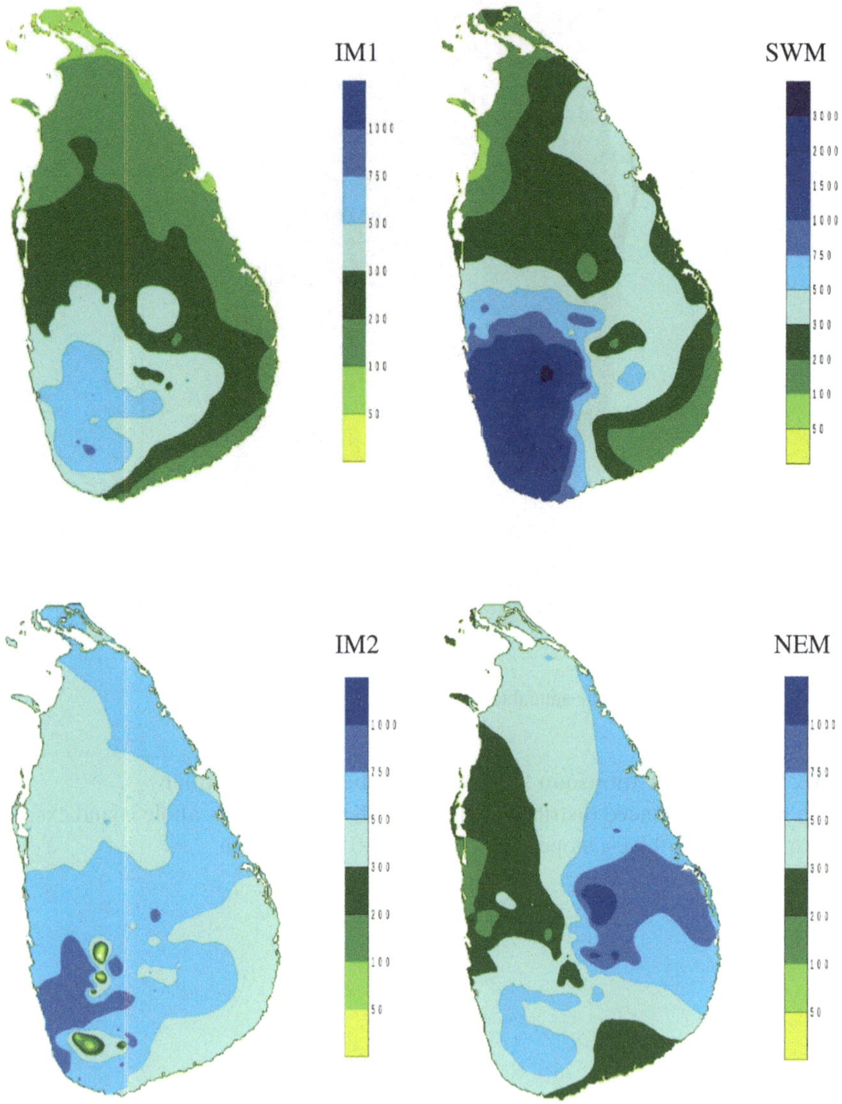

Fig. 5 Annual rainfall distribution of four seasons. *Source* Department of Meteorology, Sri Lanka

4 Terrain

4.1 Topography

The pear-shaped island has low and flat terrain where northern and coastal areas are plains and the inland in the southern and central parts are plateaus. The core area of the Central Highlands has such complex topographical features such as ridges, peaks, plateaus, basins, valleys and steep slopes. The highest one is the Pidurutalagala Mountain with 2524.13 m above sea level. The other regions are flat with an average altitude of about 150 m, except for several hills on the lowlands (Figs. 6 and 7).

4.2 Elevation Zone

From SRTMGL1 30 m DEM data, the whole land is delineated into 100 m-interval elevation zones. It can be seen from the elevation zone that the terrain of the Central Highlands in the south-central part becomes drastically higher than that of the surrounding area with an average altitude of less than 1000 m, about 30 km away from the central peak, Pidurutalagala. However, the terrain with higher than 1000 m in Sri Lanka only accounts for 4.07% of the total area. The northern and coastal areas, mostly plain with less than 100 m asl account for more than 60% of the total land area. The northern coasts are wide plain and the southern and western coasts relatively narrow. Furthermore, the area with elevation of between 100 and 200 m accounts for 19.24%, which is mainly plain (Fig. 8).

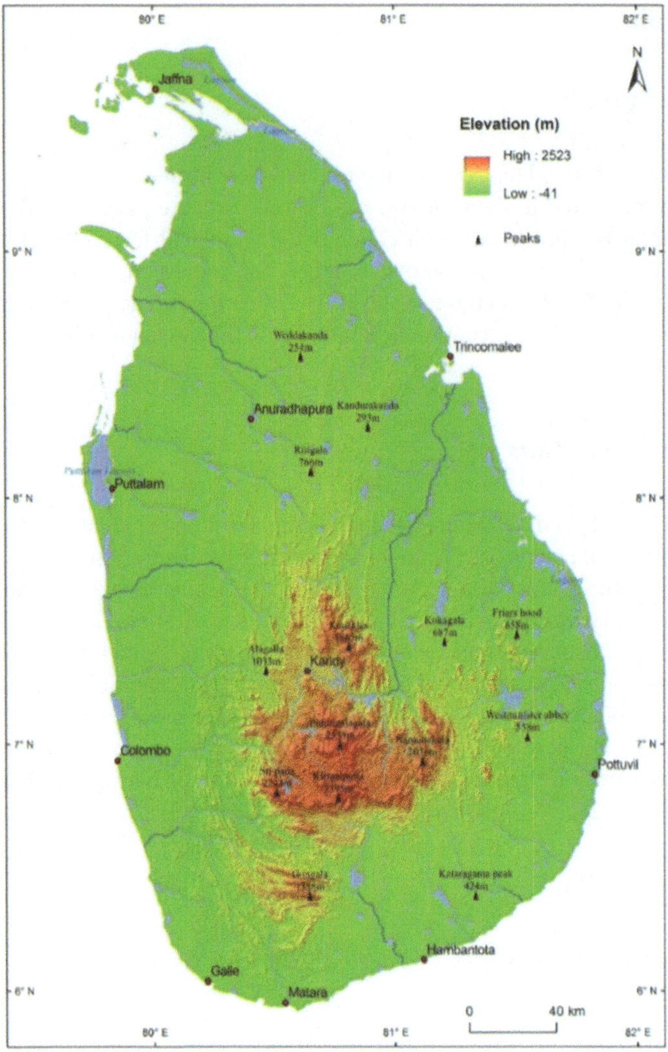

Fig. 6 Sri Lanka's terrain

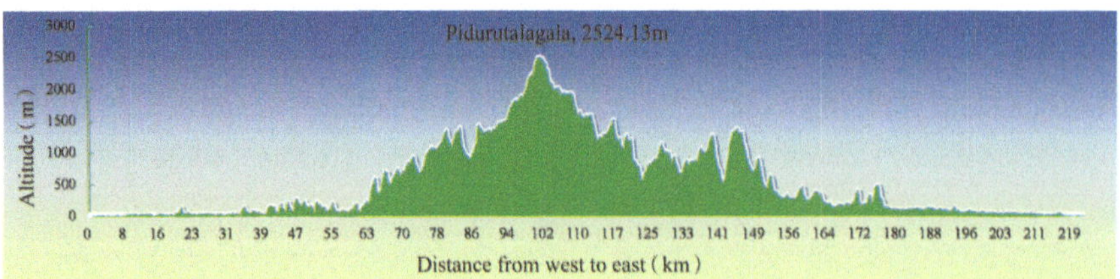

Fig. 7 Sri Lanka's topographic profile from West to East across the Pidurutalagala

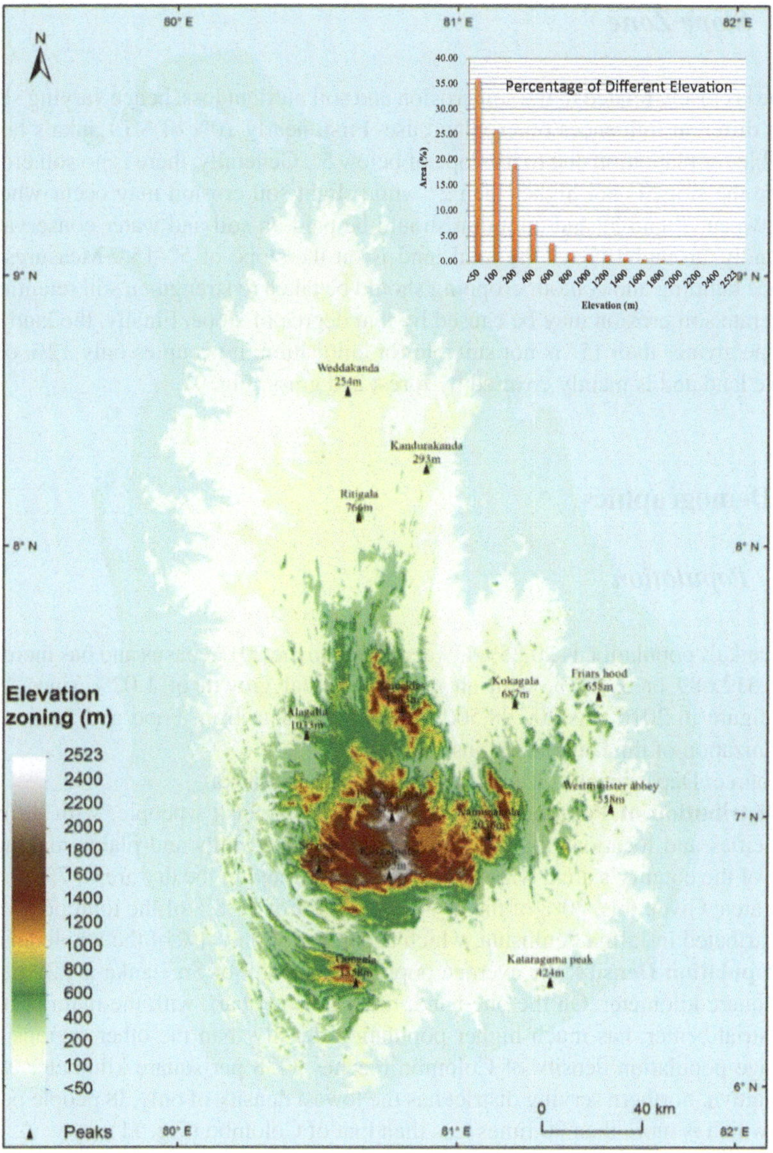

Fig. 8 Sri Lanka's elevation zone

4.3 *Slope Zone*

Slope is closely related to the soil erosion and soil nutrient loss, hence varying slopes exert different influences on cropland use. First, nearly 70% of Sri Lanka's land is suitable for plantation due to its slope of below 5°. Generally, there is no soil erosion where the slope is not higher than 2° while slight soil erosion may occur where it is between 2° and 5° and attention should be paid on soil and water conservation. Secondly, around 18% of the total land is on the slope of 5°–15°. Measures like terrace building and contour cropping should be taken to strengthen soil retention as moderate soil erosion may be caused by that degree of slope. Finally, the land with a slope greater than 15° is not suitable for cultivation, it occupies only 12% of the whole land and is mainly covered by forest and grass (Fig. 9).

5 Demographics

5.1 *Population*

Sri Lanka's population is 20,359,439 according to the 2012 census and has increased by 5,512,689 or 37.13%, with an average annual growth of 1.02% since 1981. The figure in 2016 was 20,798,500 according to data from Food and Agriculture Organization of the United Nations (Fig. 10).

Source: Department of Census and Statistics, Sri Lanka.

Distribution of Population: about 60% of Sri Lanka's people, with different ethnicities and locations, live in the southwest humid hills and plains making up 27% of the country's area, while 40% are concentrated in the dry areas, 73% of the total area. Given this, 20% of the dry-area population or 8% of the total population is distributed in Jaffna Peninsula, which accounts for only 4% of the whole land.

Population Density: the average population density of Sri Lanka is 325 people per square kilometer. On the one hand, the southwest part, with the main port and industrial center, has much higher population density than the other regions. The average population density of Colombo reaches 3438 per square kilometer, while Mullaitivu, northern serving district has the lowest density of only 38 people per sq. km, which is more than 90 times less than that of Colombo (Fig. 11).

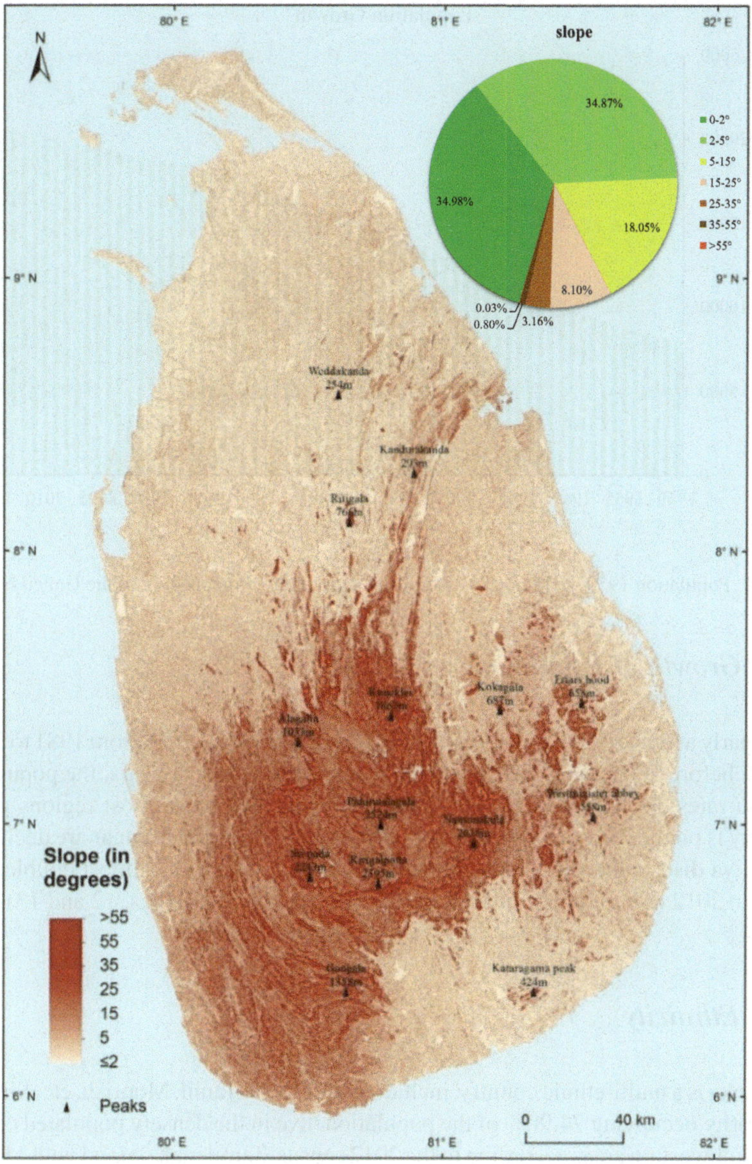

Fig. 9 Sri Lanka's slope zone

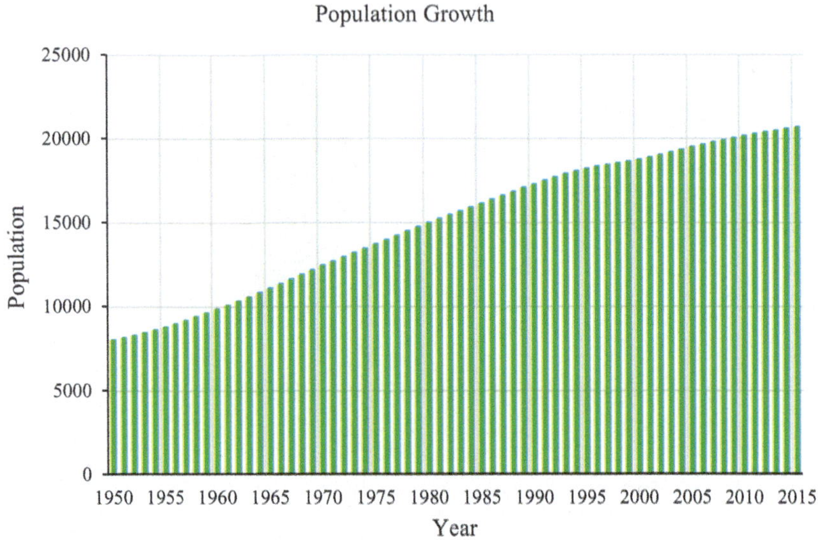

Fig. 10 Population 1950–2015. *Source* Food and Agriculture Organization of the United Nations

5.2 *Growth of Population*

The yearly average population growth rate of Sri Lanka was 1.02% from 1981 to 2012, 1.16% before 2001 and 0.75% after 2001. However, in recent years, the population growth rates are relatively lower. Although the population in most regions of the country is on the increases, the northern districts of Jaffna and Mannar are declining. Vavuniya district has the fastest growing population which has almost double from 1981 to 2012 due to immigration from the northern districts (Figs. 12 and 13).

5.3 *Ethnicity*

Sri Lanka is a multi-ethnic country, including Sinhalese, Tamil, Moorish, etc. Sinhala, a majority occupying 74.90% of the population, live in the densely populated central and southwestern areas according to the 2012 census. Tamil rank second with 15.27% where Sri Lanka's Tamil make up of 11.15% of the total population and are the largest minority mostly in the northeast. Moreover, Indian Tamil, accounting 4.12% of the total population, are distributed in the central and southern parts. Moors, 9.30%, are mainly concentrated in the southern urban areas. Additionally, there are a large number of Moors in central and eastern regions. Other ethnic groups account for

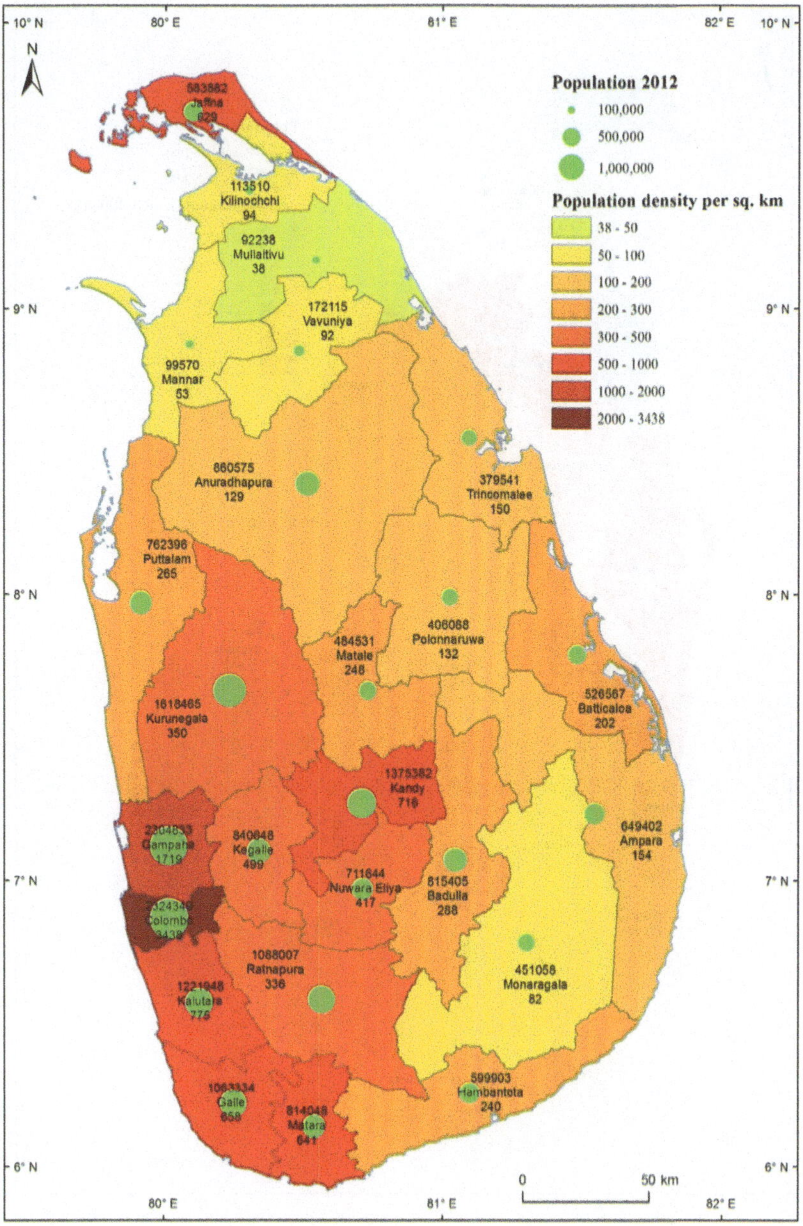

Fig. 11 Population and density

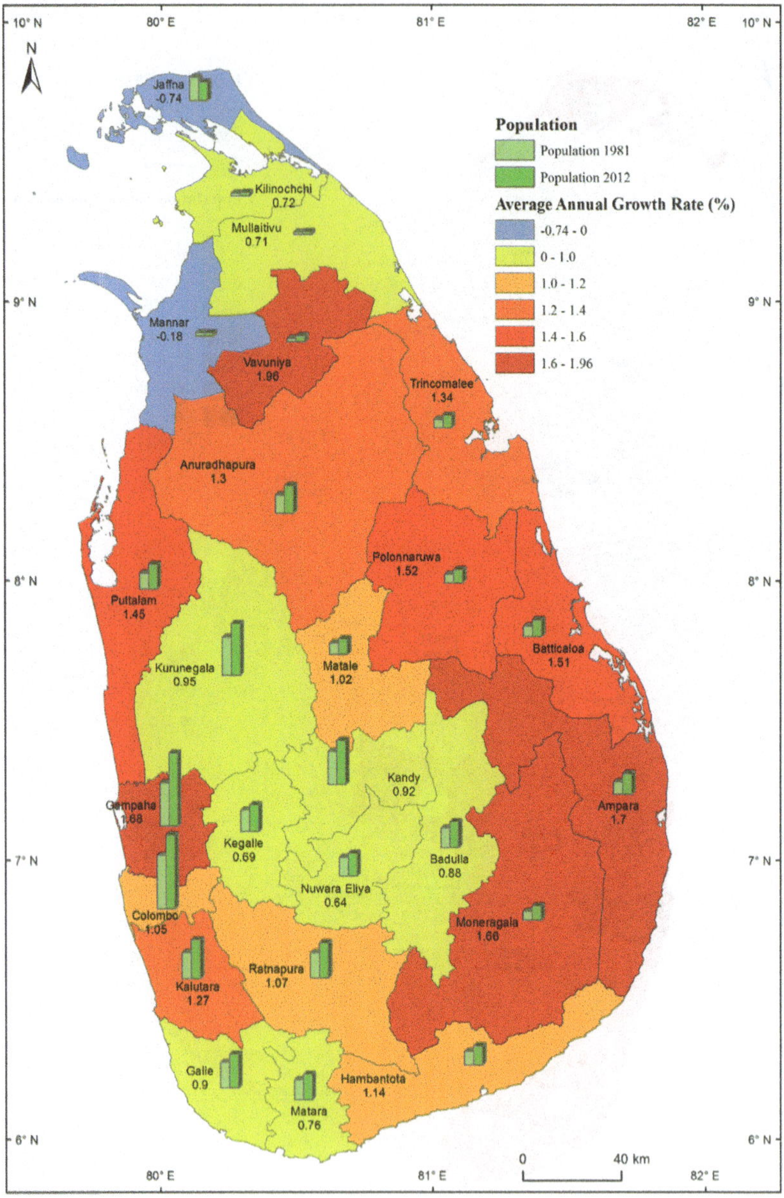

Fig. 12 Growth of population in districts 1981–2012. *Source* Department of Census and Statistics, Sri Lanka

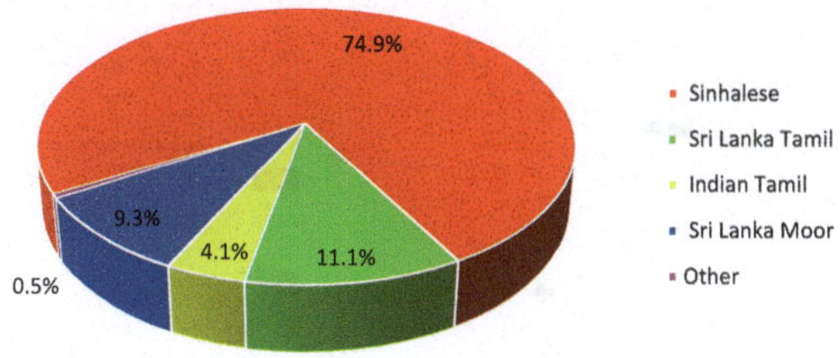

Fig. 13 Ethnicity

only 0.53%, including Malay immigrants from Southeast Asia and descendants of European colonists such as Portugal, Netherlands and UK (Fig. 14).

5.4 Religion

According to the 2012 census, Buddhists accounted for 70.10% of the population, Hindus accounted for 12.58%, Muslims accounted for 9.66%, and Christians accounted for 7.62%. Most Sinhalas are Buddhists while most Tamils are Hindus, and most Moors and Malays are Muslims. Sizable minorities of both Sinhalese and Tamils are Christians, most of whom are Roman Catholic. Most European descendants are Roman Catholics or Presbyterians while the indigenous Vedda believe in Animism and Buddhism (Figs. 15 and 16).

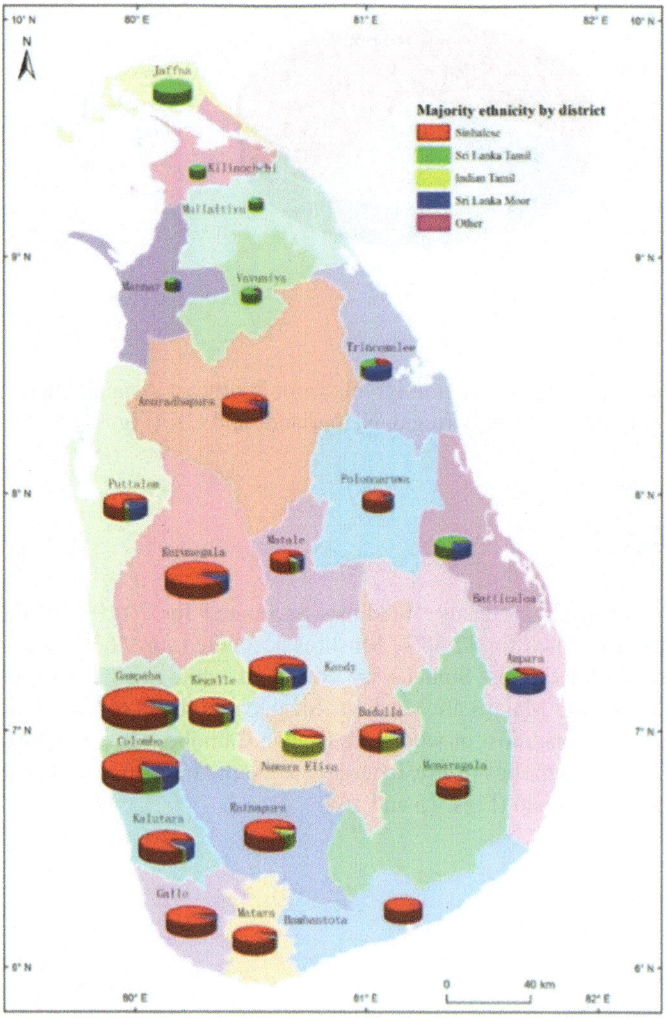

Fig. 14 Distribution of population in districts. *Source* Department of Census and Statistics, Sri Lanka

Fig. 15 Religion

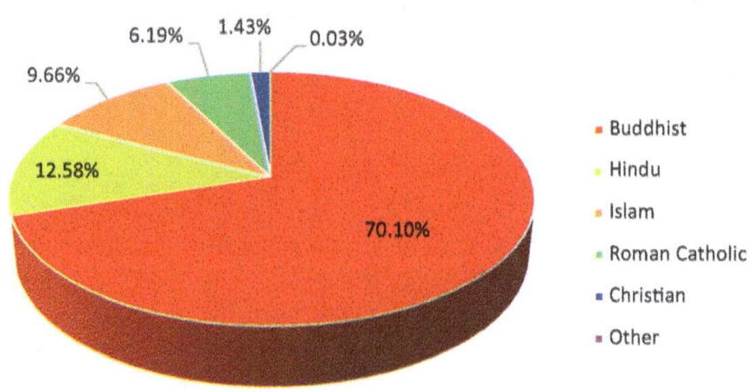

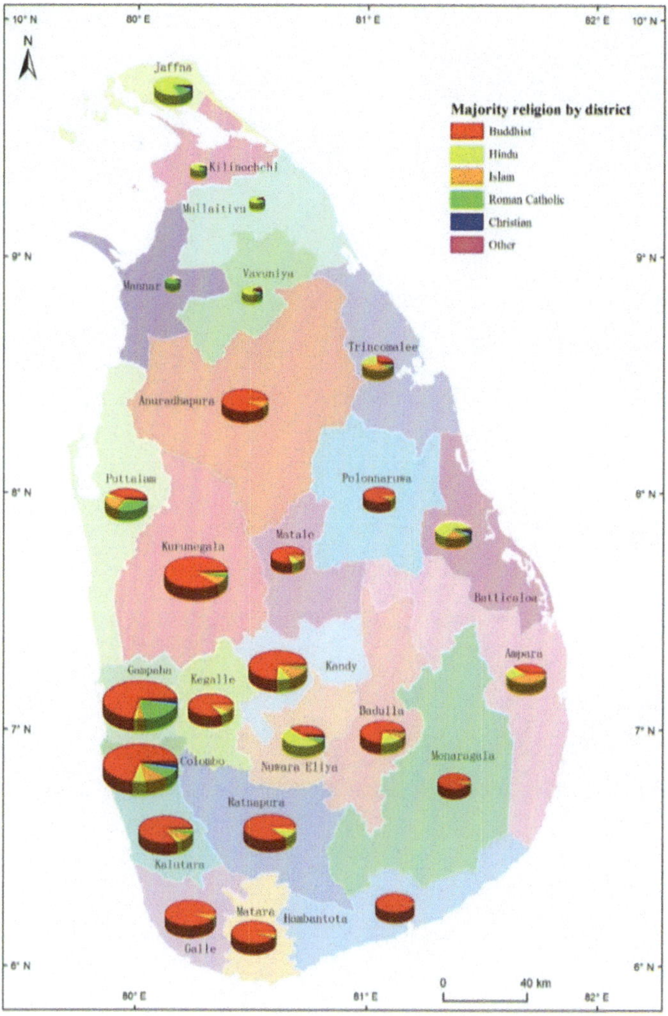

Fig. 16 Distribution of religions in districts. *Source* Department of Census and Statistics, Sri Lanka

Remote Sensing Analysis of Resources and Environment

Natural resources and environment lay a foundation for human existence and sustainable development. However, Sri Lanka has witnessed great changes in its resources and environment after decades of rapid social and economic development. Therefore, it is urgent and imperative to gain an accurate understanding of current situation and change pattern of its resources and environment for ensuring sustainable development and ecological environment protection. An irreplaceable means of large-scale dynamic monitoring in resources and environment, remote sensing can be applied in monitoring the status quo and historical changes of land, vegetation, surface water resources and marine eco-environment. By doing this, scientific data can be provided for Sri Lanka to make rational use of resources, protect ecological environment and realize sustainable economic and social development.

1 Land Use

The sustainable development of land resources is the basis of national, political, economic and cultural sustainable progress and the guarantee for the country's peace and prosperity. Data can be obtained for land resource management and policy making in Sri Lanka through land resource survey where the quantity and spatial distribution of land use types are classified

All the social and economic activities of human beings must be conducted in situ, directly or indirectly through land use thus most of the ecological and environment issues caused by human activities are associated with land use. Unfortunately, with the growing human demand in recent years, the pressure on land resources has become increasingly prominent and the ecological environment problems caused by land use are becoming more evident as well. Currently, Sri Lanka is faced with ecological environment problems related to land use to some extent such as deforestation, land degradation, lack of safe drinking water, loss of biodiversity, greenhouse gas emissions and climate change, frequent natural disasters, unsustainable management

X. Wei, *The Geography and Remote Sensing Analysis of Sri Lanka*,
https://doi.org/10.1007/978-981-33-4627-7_2

of natural resources, exploitation of coral reefs and minerals, increase of solid waste, environmental pollution. Appropriate land use policies based on land use survey and evaluation is of significance to the enduring stability and security of land ecosystem and regional ecosystem in Sri Lanka.

Land use survey and dynamic monitoring through a complete set of remote sensing techniques can offer a large number of accurate scientific data for the development and protection of land resources, macro-economic decision-making and ecological environment protection. It also plays an important role in such fields as land resource planning and governance, large engineering construction, major natural disaster emergency and disaster relief.

The land resources survey in Sri Lanka is mainly based on medium-resolution remote sensing images, data editing and validation are supplemented by ground investigation and Google Earth high-resolution remote sensing images. The medium-resolution images include a total of 11 Landsat-8 images (15 m resolution) without or with little cloud across the country, mainly in 2016 and 2017.

The establishment of land use classification system provides a basis of land use remote sensing mapping, monitoring, and analysis. This classification system is formulated after full investigation of land resource characteristics and the analysis of remote sensing images (Table 1).

Table 1 Land use classification system of remote sensing survey

Code	Types	Description
10	Cropland	Cultivated lands for crops: the lands for paddy, vegetable, grass, fruit trees or other economic plants; the irrigated or rainfed land; tea or coffee garden or other plantations for economic shrub
20	Forest land	The arbor crown density of more than 30%: the forests with deciduous hardwood, evergreen hardwood, deciduous conifer or evergreen conifer; mixed forest or sparse forest with crown density of 10–30%
30	Grassland	The natural herbaceous vegetation density of more than 10%: grassland, meadow, savanna, desert steppe or urban artificial grassland
40	Shrubland	The shrub density higher than 30%: upland shrub; deciduous or evergreen shrub; desert shrub with desertification of over 10%
50	Wetland	Located in the border zone between dry land and water, it features shallow water or fairly wet soil mostly with helophyte or hygrophyte. E.g. lake swamp, river flooding wetlands, forest or shrub wetland, peat bog, mangrove and salt marsh
60	Water bodies	All covered by liquid water: rivers, lakes, reservoirs, ponds, etc.
70	Built-up Land	The areas formed by artificial construction activities: various residential areas such as cities and towns; industrial and mining sites; transportation facilities. Also, the green areas and waters inside the built-up land is excluded
80	Bare land	The land with only soil, rock or gravel on the surface and basically no vegetation

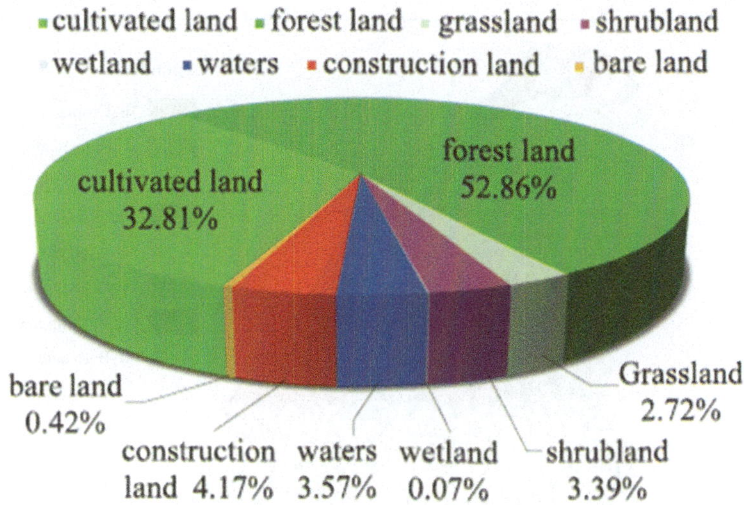

Fig. 1 The proportion of land use

1.1 *Land Use Status in Sri Lanka*

The country has an area of 66,185.9 km^2 according to remote sensing monitoring for land use in 2017, including 8 types land use land cover such as cropland, forest, grassland, shrubland, wetland, water bodies, built-up land and bare land. In light of its proportion, forest land ranks first, accounting for 52.86%, followed by cropland, 32.81%, while built-up land is widely distributed, totaling 4.17%. There are many water bodies such as rivers, lakes, reservoirs and ponds with a relative high coverage of 3.57%. Due to abundant rainfall suitable for the growth of trees and crops, the proportion of forest land and crop land in the land use is relatively large, while shrubland, grassland and bare land are less with 3.39%, 2.72% and 0.42% respectively and wetland is the lowest, making up 0.07% (Figs. 1 and 2).

1.2 *Distribution Characteristics of Land Use by Provinces*

Both cropland and forest land dominate the land use structure of different provinces and is similar to that of the whole country, which account for a large proportion of the total area of each province ranging from 75 to 95%. However, there are huge differences in their proportions. For example, the proportion of forest land area in Sabaragamuwa is fairly high—8.75 times of that of the second largest land use, and the proportion of cropland in Eastern and North Central provinces are the largest—slightly larger than that of forest land. In addition to cropland and forest, the proportion of other land types in each province is relatively small and their proportions varies in different provinces. Moreover, the proportion of built-up land area differs significantly among provinces, with the highest coverage in Western province of

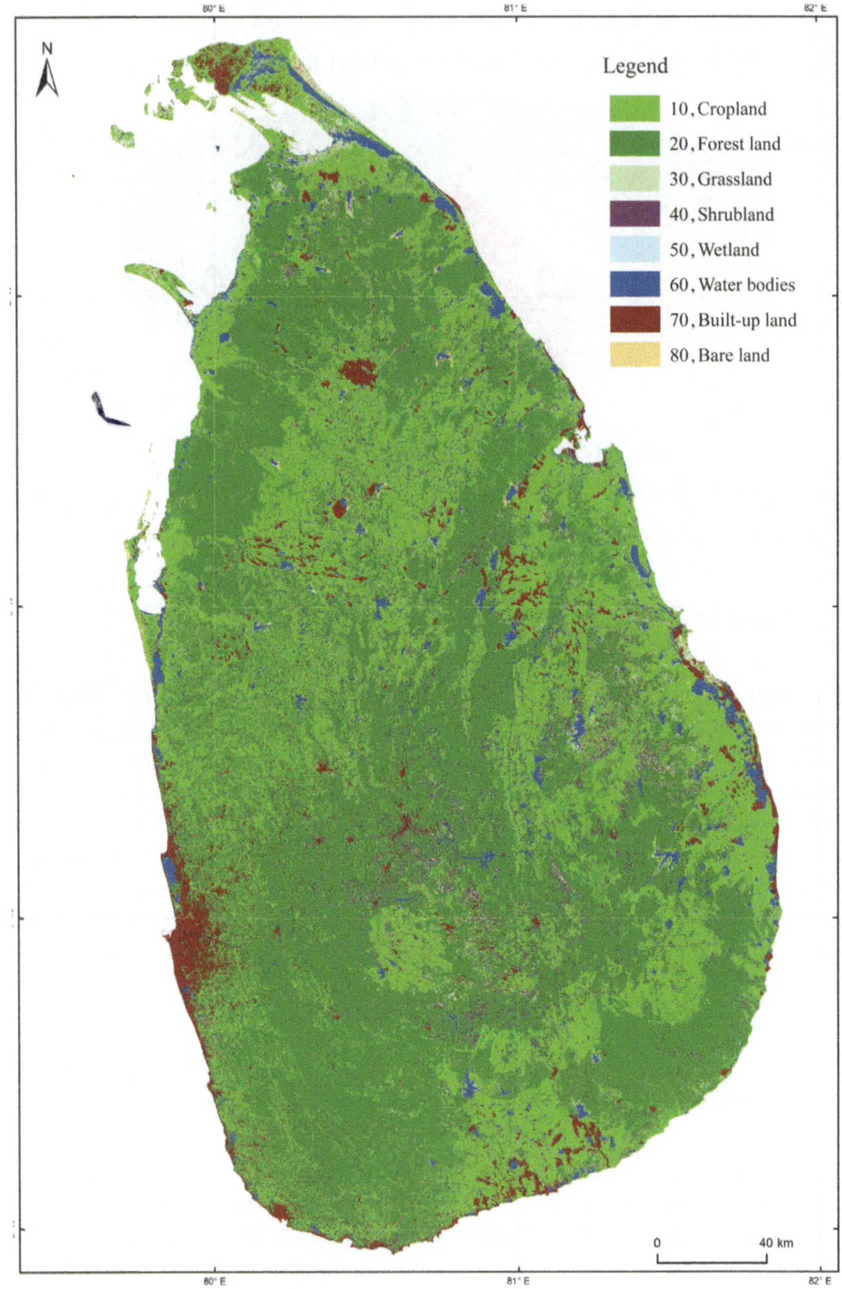

Fig. 2 Land classification by remote sensing

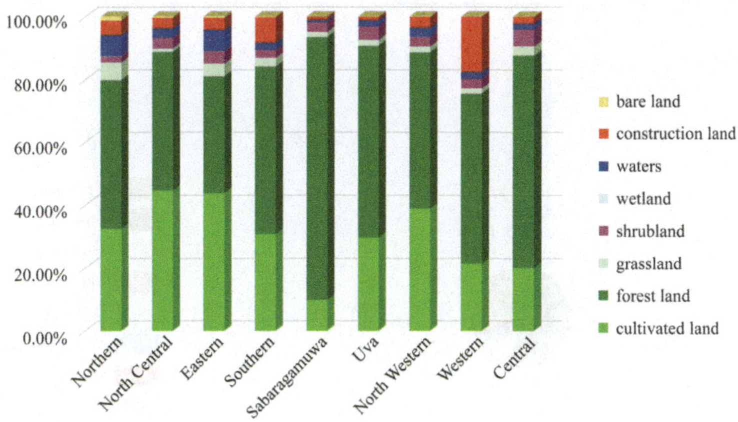

Fig. 3 The proportion of land by provinces

17.56% and the lowest two in Uva and Sabaragamuwa provinces with 0.95% and 0.98% respectively—17 times less than that of Western Province. The proportion of waters area in different provinces is also varies greatly. The relatively larger ones, Eastern (6.77%) and Northern (6.67%), are both almost more than 8 times of the smallest one, Sabaragamuwa of 0.78% (Figs. 3 and 4 and Tables 2 and 3).

Distribution Characteristics of Cropland by Provinces

As the most important type of land use for population capacity, cropland is widely distributed in every province. North Central Province has the largest part, reaching 21.80% of the total cropland area, followed by Eastern, 19.73% with the two accounting for more than 40% of the national total. These areas are not only enormous but also centralized. The cropland area in Sabaragamuwa Province is the smallest and decentralized contributing 2.19% to the national total cropland.

The cropland in Sabaragamuwa is the smallest and the lowest proportion in the province covering only 9.59%. However, the cropland proportions of the rest provinces are all about 20% or higher. Specifically, North Central and Eastern provinces have the largest proportion of cropland area covering more than 40% of their respective provincial areas (Figs. 5 and 6).

Distribution Characteristics of Forest Land by Provinces

The largest land use type in Sri Lanka, forests is widely distributed but relatively small in Western and Southern provinces, totaling 5.80% and 8.45% of the national forests area respectively. The forest land areas of the other seven provinces are relatively large with smallest differences amongst themselves each ranging from 10% to 15% of the whole forest land area. Contrastingly, their proportion within provinces varies significantly. Sabaragamuwa Province, ranking seventh in contribution to the total forest land area, has the highest proportion of forest land area of 83.88%, while Eastern Province has the lowest proportion of 37.36% (Figs. 7 and 8).

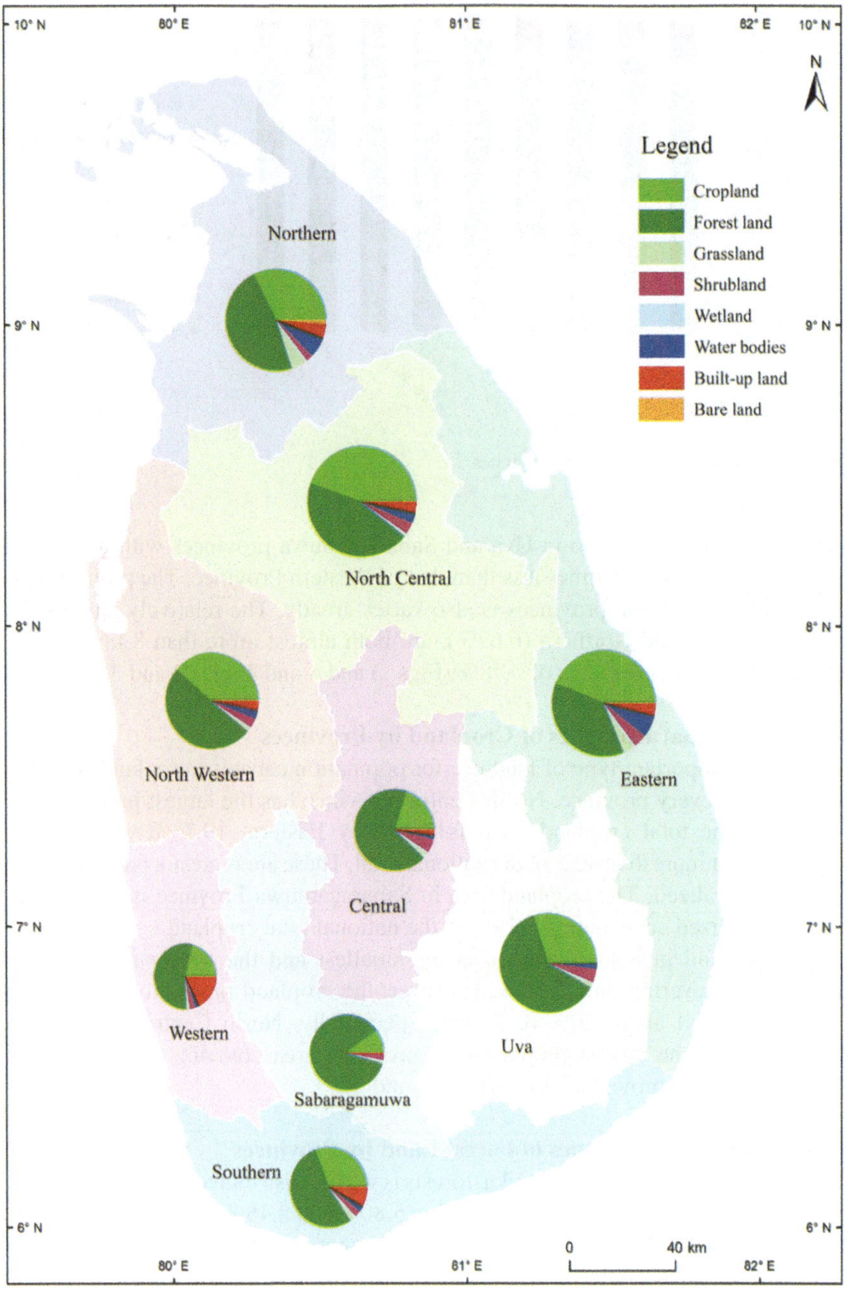

Fig. 4 The characteristics of land use structure by provinces

Table 2 Area of different land use by provinces (sq. km)

Province	Cropland	Forest land	Grassland	Shrubland	Wetland	Waters	Built-up land	Bare land	Total
Northern	2922.8	4282.5	514.4	169.4	14.4	602.3	411.0	114.2	9031.0
North Central	4733.2	4714.0	113.5	356.5	5.4	330.8	345.8	37.8	10637.0
Eastern	4282.9	3666.9	401.3	381.9	9.3	664.9	369.1	40.0	9816.2
Southern	1695.1	2954.6	151.3	139.9	2.7	128.6	443.1	11.8	5527.2
Sabaragamuwa	474.6	4152.8	81.2	153.3	0.2	38.8	48.5	1.4	4950.9
Uva	2550.6	5274.2	154.6	374.8	1.7	169.1	81.7	13.9	8620.6
North Western	3113.8	4017.1	149.1	244.1	7.1	233.5	266.4	12.8	8043.8
Western	796.0	2029.9	61.8	113.4	1.2	83.4	657.2	0.5	3743.4
Central	1143.4	3888.5	165.8	308.2	0.7	105.5	122.0	3.5	5737.6

Table 3 The proportion of land use by provinces (%)

Province	Cropland	Forest land	Grassland	Shrubland	Wetland	Waters	Built-up land	Bare land	Total
Northern	32.36	47.42	5.70	1.88	0.16	6.67	4.55	1.26	100.00
North Central	44.50	44.32	1.07	3.35	0.05	3.11	3.25	0.36	100.00
Eastern	43.63	37.36	4.09	3.89	0.09	6.77	3.76	0.41	100.00
Southern	30.67	53.46	2.74	2.53	0.05	2.33	8.02	0.21	100.00
Sabaragamuwa	9.59	83.88	1.64	3.10	0.00	0.78	0.98	0.03	100.00
Uva	29.59	61.18	1.79	4.35	0.02	1.96	0.95	0.16	100.00
North Western	38.71	49.94	1.85	3.03	0.09	2.90	3.31	0.16	100.00
Western	21.26	54.22	1.65	3.03	0.03	2.23	17.56	0.01	100.00
Central	19.93	67.77	2.89	5.37	0.01	1.84	2.13	0.06	100.00

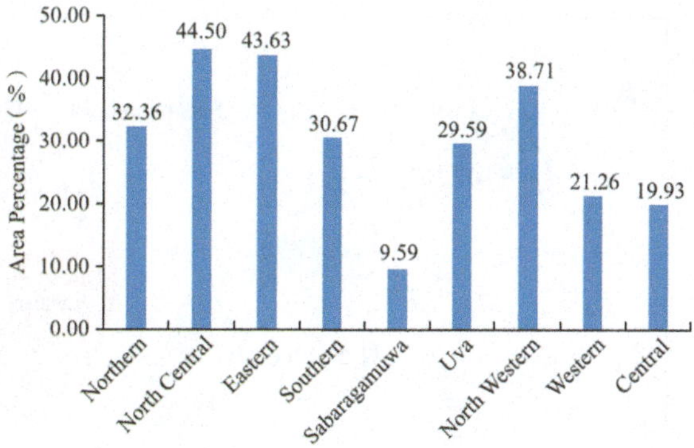

Fig. 5 Proportion of cropland by provinces

Distribution Characteristics of Grassland by Provinces

Sri Lanka has a small grassland area, covering only 2.72% of the country's land. More than half of the country's grassland area is distributed in Northern and Eastern provinces, while the rest parts of the country have relatively small grassland area and Western Province has the least coverage, accounting for 3.45% of the country's grassland area. The proportion of grassland area in each province is very small, such that the highest coverage in Northern Province covers only 5.70% of the provincial area, followed by 4.09% in Eastern Province, 2.74% in Southern Province which is slightly higher than the national average while other six provinces have lower than the national average (Figs. 9 and 10).

Distribution Characteristics of Shrubland by Provinces

Shrubland, small but dispersed, is mainly distributed in the transition zone of forest land and grassland. Geographically, only 3.39% of the national land use is shrubland with nearly half of national total in Eastern, Uva and North Central provinces while the Western Province has he least, contributing to 5.06% of the national total shrubland. The proportion of shrubland area in each province is small, with the highest proportion of 5.37% in Central Province (Figs. 11 and 12).

Distribution Characteristics of Waters by Provinces

Water area accounts for 3.57% of the country's land, which is higher than the global water area rate of 2.73% and the Asian rate of 2.79%. Sri Lanka houses abundant water resources. Rivers, lakes, reservoirs and ponds are widely distributed throughout the country, particularly in such arid regions as Eastern, Northern and North Central provinces totaling 68% of the national water area (28.21%, 25.55% and 14.04% respectively). Given less rainfall in those arid regions, people always attach great importance to the construction of reservoirs and ponds to reserve the abundant water resources in the rainy season for the dry season. By doing this, the agricultural irrigation and living water can be guaranteed. The humid southwest regions have relatively small water area and much less reservoirs and ponds due to its abundant rainfall.

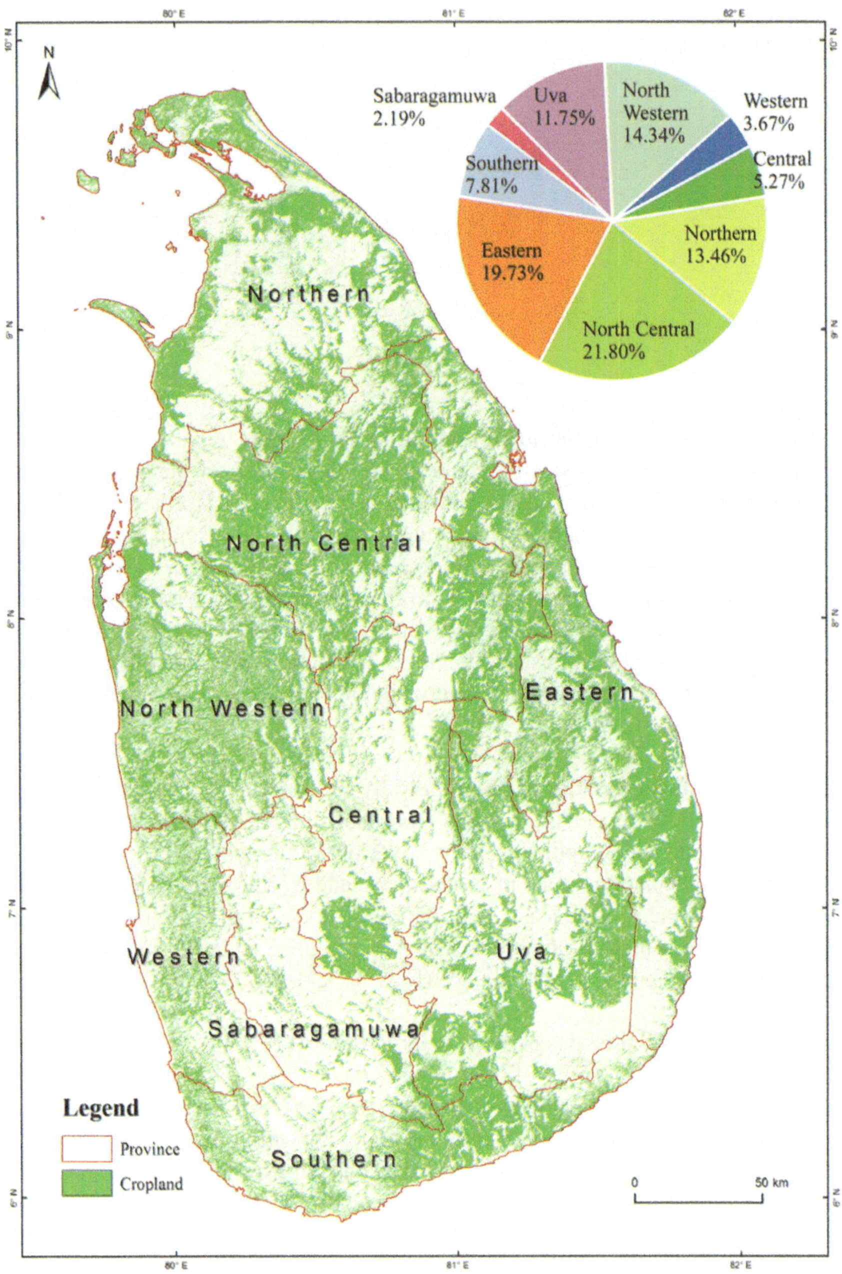

Fig. 6 Distribution characteristics of cropland by provinces

Fig. 7 Proportion of forest land area by provinces

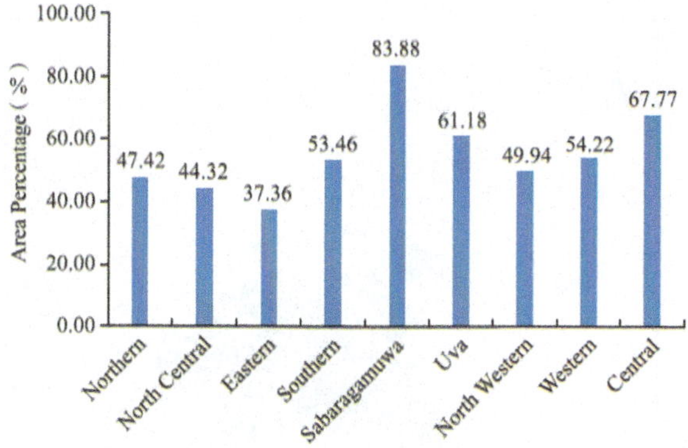

Geographically, the higher water area rates are 6.77% in Eastern Province, 6.67% in Northern Province and 3.11% in North Central Province. Among them, the Northern Province has the lowest water coverage and mainly has fresh water resources for agricultural irrigation, while the other two are dominated by lagoons (Figs. 13 and 14).

Distribution Characteristics of Built-Up Land by Provinces
With extensive distribution and regional differences, built-up land is the completely human changed the original natural land cover to artificial surface. It makes up 4.17% of the country's land and vary significantly throughout provinces. Western Province has the largest built-up area, accounting for 23.95% of the country's built-up land; the following by Southern Province, accounting for 16.14%. Both of these provinces are the most developed regions in Sri Lanka and consequently with more urban built-up land. Also, there are significantly more built-up land areas in Northern and North Central provinces and here are more ancient cities and ancient architectures found. Eastern as well has more built-up land, mostly along its eastern coastal areas. Sabaragamuwa and Uva provinces have the least built-up area mainly sparsely scattered rural residential sites.

Across the country, Western Province has a much higher proportion of built-up land than other provinces, making up 17.56% of the total provincial area mainly distributed in city of Colombo and surrounding areas. Southern Province ranks second, with built-up land accounting for 8.02% of the provincial land, mostly in and around the southwest coastal cities. Northern Province ranks third with 4.55%, mainly in and around the provincial capital city of Jaffna. Sabaragamuwa and Uva have built-up area less than 1% of their respective provincial areas (Figs. 15 and 16).

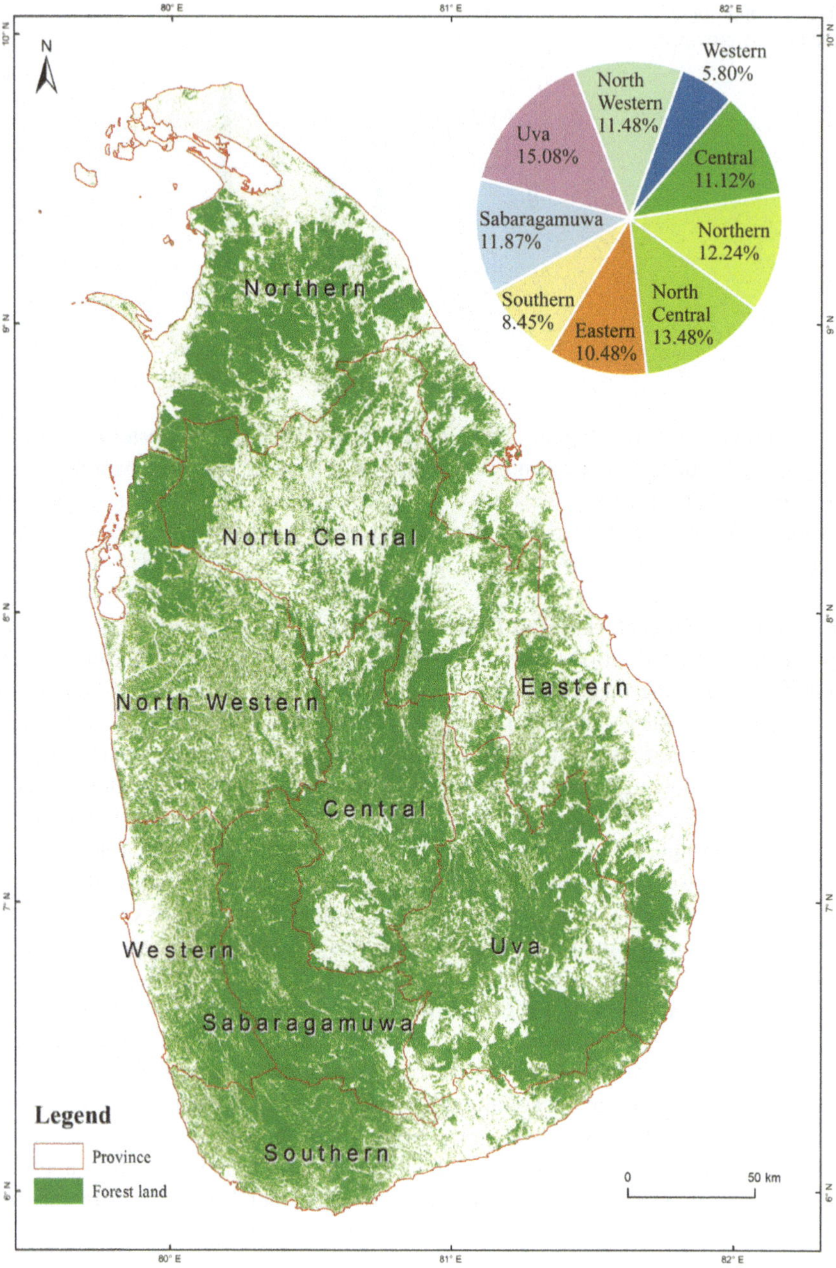

Fig. 8 Distribution of forest land by provinces

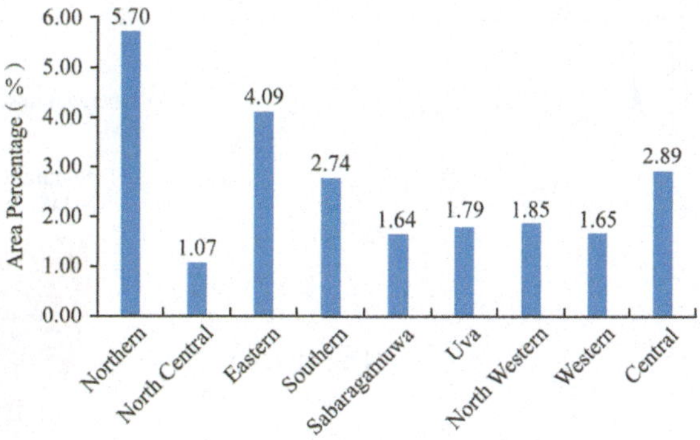

Fig. 9 Proportion of grassland area by provinces

1.3 Distribution Characteristics of Land Use Under Different Elevations

Elevation is one of the important natural factors that affect land use. Table 4 lists the proportion of various land use area in different elevation zones (EZs), where there are notable differences. The cropland is concentrated in the area with the elevation below 200 m, taking up 93.17% of the country's cropland area. Being low and flat, it produces the majority of the national crops and consists majorly of paddy fields and dry land. In the structure of land use types below 200 m, the proportion of cropland is the most common, accounting for more than one third of the total area. Moreover, in the mountainous zones with elevation ranging from 1000 to 2000 m, cropland is quite common, mainly consisting tea plantation. However, cropland in the other elevation zones is rare and relatively small. Forests, the biggest land use in Sri Lanka, has 72.11% its proportion distributed below the elevation of 200 m although it's widely spread and dominating in most elevation zones. Of the grassland scattered in all EZs, 82.15% is located below the elevation of 200 m. Nevertheless, the proportion of grassland in the zone above 1000 m is generally higher than that below 1000 m. Similarly, 71.91% of shrubland distributed in all EZs locates the zone below 200 m, but the highest proportion is concentrated in the zone within 500–1500 m. The wetland is relatively small and mainly distributed below the elevation of 200 m. 95.62% of the water area is below the elevation of 200 m, mainly rivers, reservoirs and ponds used for agriculture and living. In addition, there are a small number of water areas in the range of 200 m to 500 m, mainly mountain reservoirs, while only 1.17% of the total water area is in the range of 500 m above the elevation. As for the built-up land, 93.55% is distributed below the elevation of 200 m. Except for the elevation zone below 100 m having a higher proportion of the built-up land, it occupies around 2% of each of the other elevation zones. That shows that the natural environment in different elevation zones in Sri Lanka is suitable for living. The bare land in Sri Lanka is small and mainly distributed in the range below 200 m elevation (Fig. 17 and Table 5).

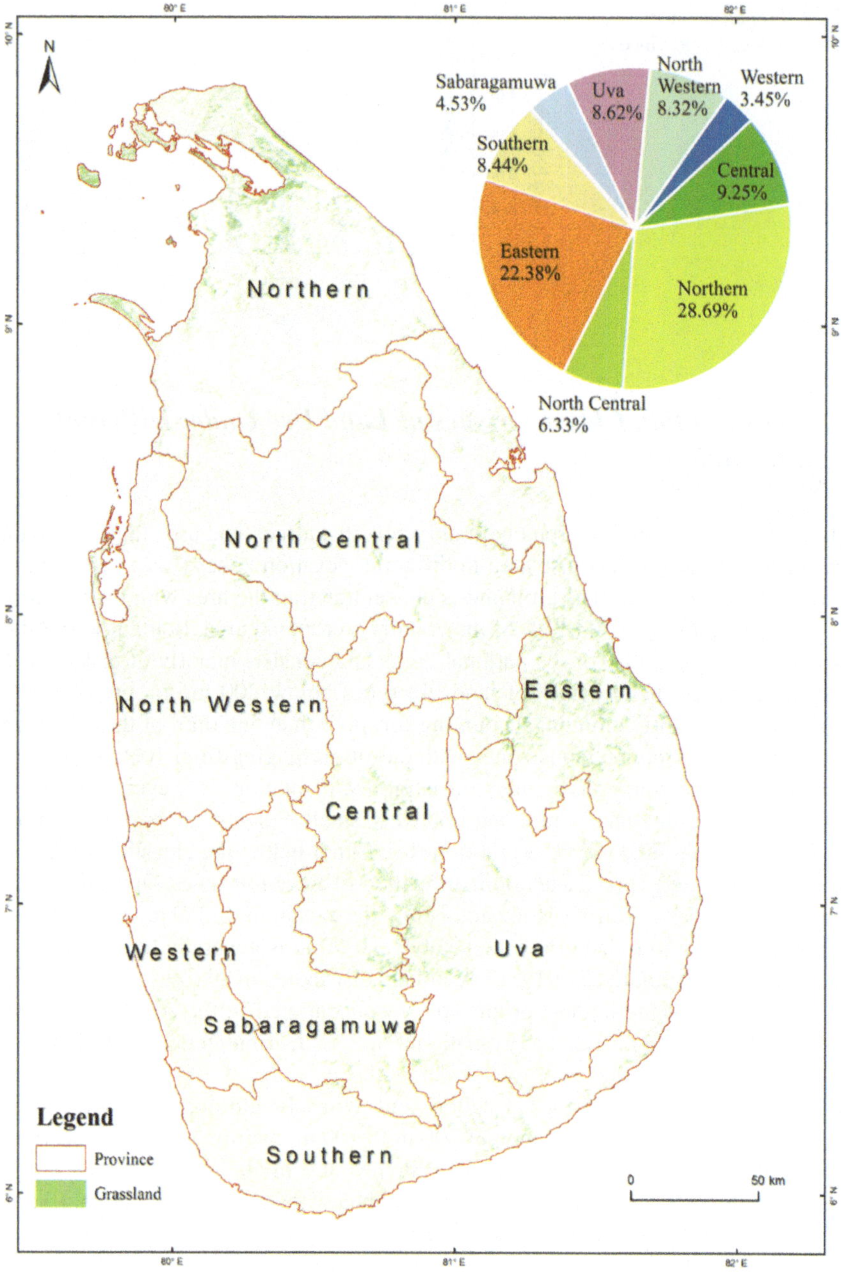

Fig. 10 Distribution characteristics of grassland by provinces

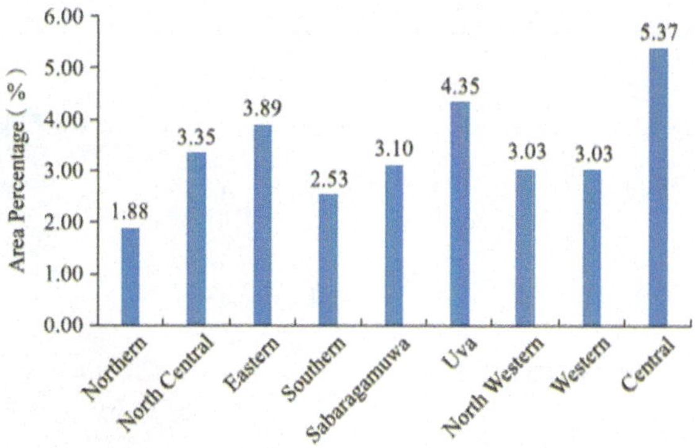

Fig. 11 Proportion of shrubland by provinces

1.4 Distribution Characteristics of Land Use Under Different Slope Gradient

Slope is another important topographic factor that affects land use, especially for cropland and built-up land. In some cases, it may restrict their development and utilization. Table 6 shows the proportion of different land use under different slope positions where all types of land are basically distributed in the range of slope below 15°. Specifically, most cropland, wetland, waters, built-up land and bare land are located in the slope zone of 0°–5° and rarely in that of 5°–15° and most forest land, grassland and shrubland are situated in the range of 0°–5° and some in 5°–15° and 15°–25° gradients (Fig. 18).

The land use varies in different slope gradient. Notably, cultivated land is the largest in 0°–2° slope gradient, followed by forest land. Waters is the third largest with 6.10% coverage within this slope, higher than the national figure of 3.57% across all the slope grades. Moreover, the proportion of forest land, grassland and shrubland nearly proportionate in each slope gradient with the forest land dominating all grades especially slopes above 5° (Table 7).

2 Vegetation Cover

Vegetation, an important part of the ecological environment, is the general name of all plant communities on the earth's surface. Species, quantity and distribution are important indicators to evaluate the regional ecological environment. Prerequisite to the ecological environment protection is first the protection of the surface vegetation. For this cause, the analysis of the current vegetation status and its change characteristics in the past two decades is of great significance and the basis of the ecological environment protection in Sri Lanka.

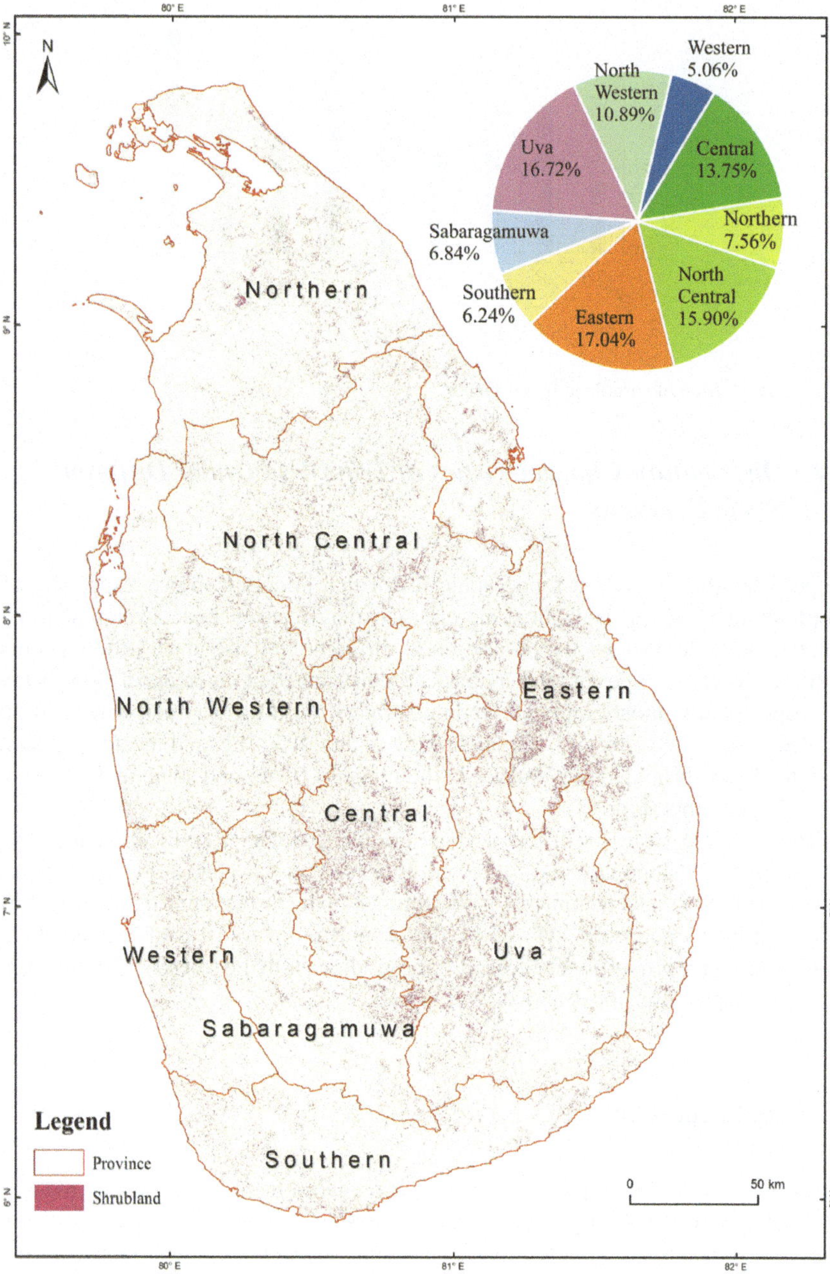

Fig. 12 Distribution characteristics of shrubland by provinces

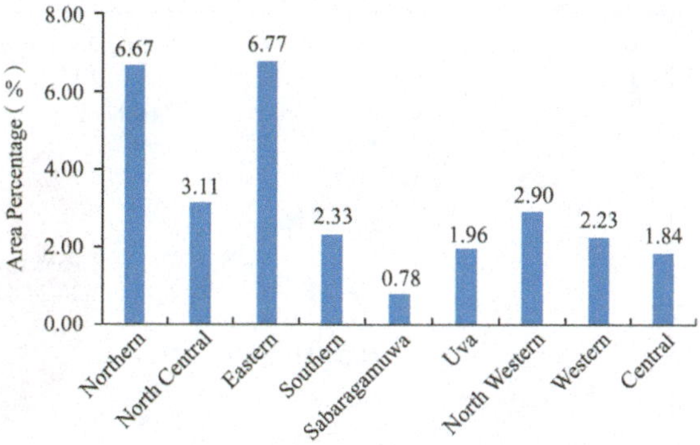

Fig. 13 Proportion of waters by provinces

A large number of studies have proved that the vegetation indexes calculated from the spectral information of vegetation can accurately reflect the changes of vegetation growth, biophysical and biochemical characteristics and ecological parameters. Therefore, it is common to use remote sensing technology to analyze the dynamic changes of vegetation Normalized Difference Vegetation Index (NDVI), as one of the best indicators of vegetation growth and vegetation coverage, and can reflect the vegetation coverage information of the study area on a large scale in time and space widely used in the sectors of ecology, environment and agriculture. The long-term and low spatial resolution remote sensing data are generally used in the study of the dynamic changes of large-scale vegetation cover. The Vegetation Index data utilized here are from the 2000–2018 MOD13Q1-NDVI time series data acquired from United States Geological Survey (http://glovis.usgs.gov/), with a spatial resolution of 250 m and a temporal resolution of 16 days.

The greatest advantage of MOD13Q1-NDVI data in vegetation cover research lies in its long-term and continuity. A total of 868 images (2000–2018) from MOD13Q1-NDVI data including the tiles of h25v08 and h26v08 have been collected for the vegetation cover monitoring in Sri Lanka.

The indicators, methods and characteristics in the following analysis are as follows:

- Monthly Vegetation Index

It is composed of the maximum value in the month and can be used as the main parameter to reflect the vegetation growth in the month. The calculation formula is:

$$NDVI_i = Max(NDVI_{ij})$$

$NDVI_i$ is the NDVI value of the i-th month and $NDVI_{ij}$ is the maximum composite value of the j-th day of the i-th month.

- Monthly Average Vegetation Index

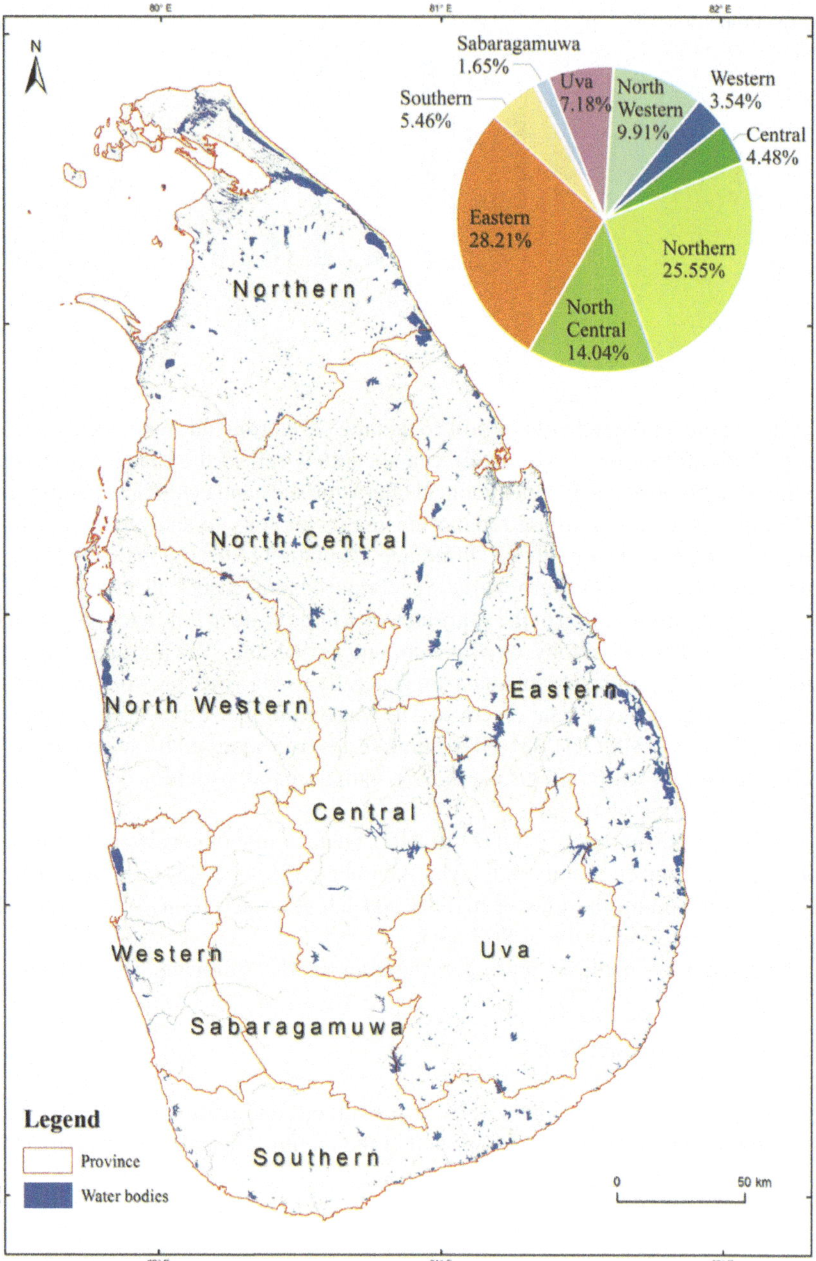

Fig. 14 Distribution characteristics of waters by provinces

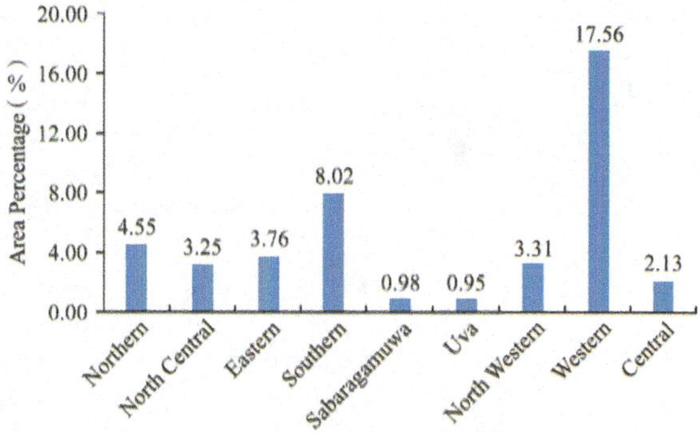

Fig. 15 Proportion of built-up land by provinces

It can be used as a parameter to reflect the long-term characteristics of vegetation. It is the average value of the Monthly Maximum Vegetation Index expressed as:

$$\overline{NDVI_i} = \left(\sum_{j=1}^{N} NDVI_{ij}\right) \Big/ N \quad i = 1, 2, \ldots 12; \;\; j = 1, 2, \ldots N$$

$\overline{NDVI_i}$ is for the average Vegetation Index of the i-th month, N for the number of years, i for the month and j for the year. $NDVI_{ij}$ is the NDVI of the i-th month of the j-th year.

- Yearly Average Vegetation Index

As the average value of the vegetation index in each month of a year, it reflects the average situation of the vegetation in the year. Besides, the analysis of the annual change of vegetation based on this Index gives a good reflection of the annual change trend of the Vegetation Index calculated as:

$$NDVI_y = \left(\sum_{j=1}^{N} \sum_{i=1}^{M} NDVI_{ij}\right) \Big/ M \quad i = 1, 2, \ldots 12; \;\; j = 1, 2, \ldots N$$

$NDVI_y$ represents the average value of monthly NDVI in a year and $NDVI_{ij}$ is the NDVI in the i-th month of the j-th year, N for the number of years and M for the number of months.

- Image Differencing Method

It is the subtraction of the NDVI values of all grids in the previous period from that of the same position in the later period, which can show the annual change characteristics of vegetation cover in the study area for two years. The calculation formula is:

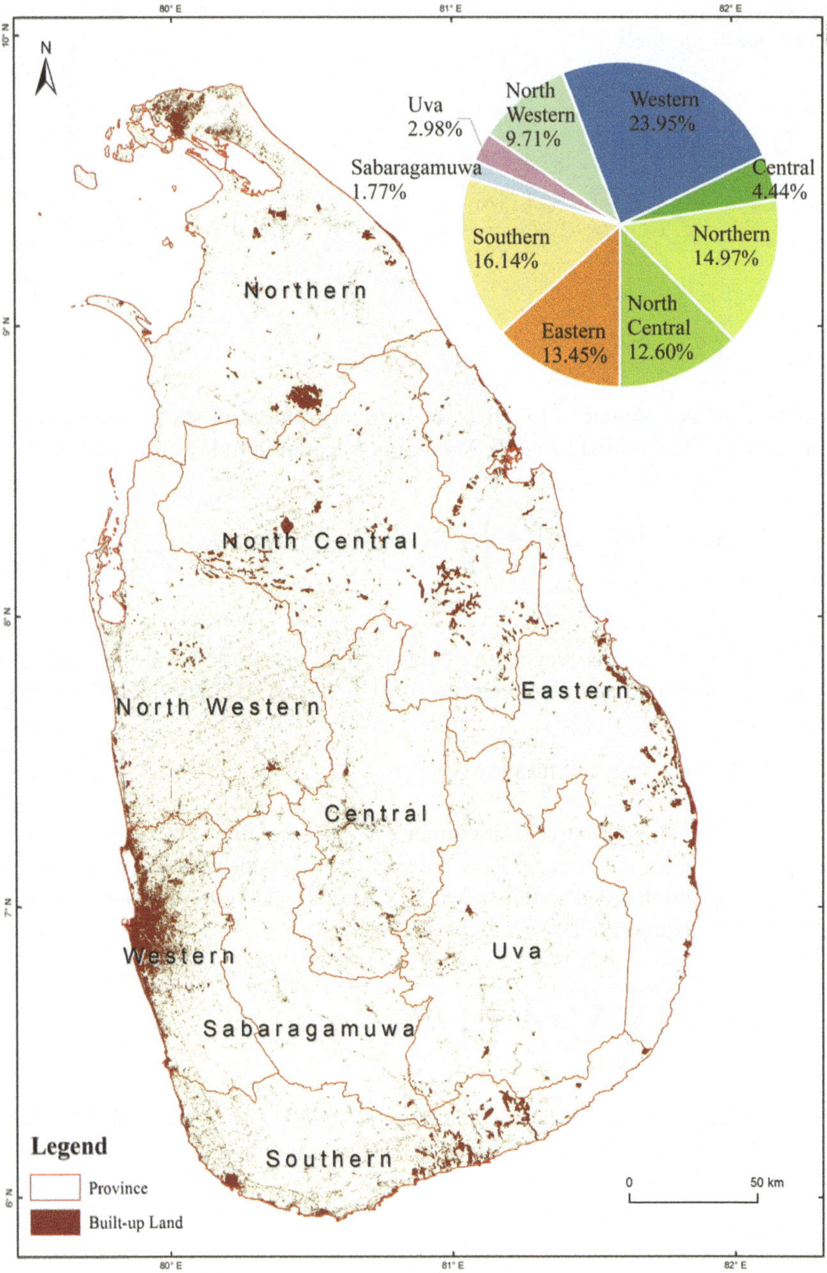

Fig. 16 Distribution characteristics of built-up land by provinces

Table 4 Proportion of different land use by elevation zones (%)

Elevation zone (m)	Cropland	Forest land	Grassland	Shrubland	Wetland	Waters	Built-up land	Bare land
0–100	73.78	51.17	74.41	49.82	88.84	84.61	85.34	93.98
100–200	19.39	20.94	7.75	22.10	10.30	11.01	8.22	3.94
200–300	1.58	7.63	1.56	5.02	0.43	1.74	0.70	0.07
300–400	0.38	4.25	0.83	2.59	0.03	0.22	0.46	0.01
400–500	0.34	3.21	0.97	2.75	0.12	1.25	0.97	1.47
500–600	0.31	2.85	0.82	3.07	0.05	0.13	0.80	0.01
600–700	0.16	1.75	0.69	1.93	0.01	0.27	0.33	0.15
700–800	0.20	1.39	0.69	1.67	0.01	0.04	0.17	0.10
800–900	0.22	1.18	0.87	1.87	0.00	0.02	0.14	0.00
900–1,000	0.29	0.94	1.09	1.71	0.00	0.02	0.15	0.01
1,000–1,100	0.37	0.83	1.28	1.68	0.02	0.16	0.25	0.06
1,100–1,200	0.45	0.72	1.32	1.64	0.02	0.36	0.37	0.02
1,200–1,300	0.57	0.61	1.46	1.44	0.02	0.02	0.54	0.02
1,400–1,500	0.70	0.40	1.28	0.96	0.01	0.01	0.34	0.01
1,500–1,600	0.52	0.27	0.87	0.53	0.01	0.00	0.18	0.01
1,600–1,700	0.31	0.26	0.85	0.40	0.00	0.00	0.14	0.01
1,700–1,800	0.18	0.26	0.65	0.28	0.00	0.00	0.11	0.00
1,800–1,900	0.08	0.25	0.46	0.15	0.00	0.00	0.09	0.01
1,900–2,000	0.06	0.31	0.51	0.12	0.03	0.06	0.18	0.02
>2000	0.11	0.77	1.64	0.27	0.11	0.07	0.53	0.10
Total	100.00	100.00	100.00	100.00	100.00	100.00	100.00	100.00

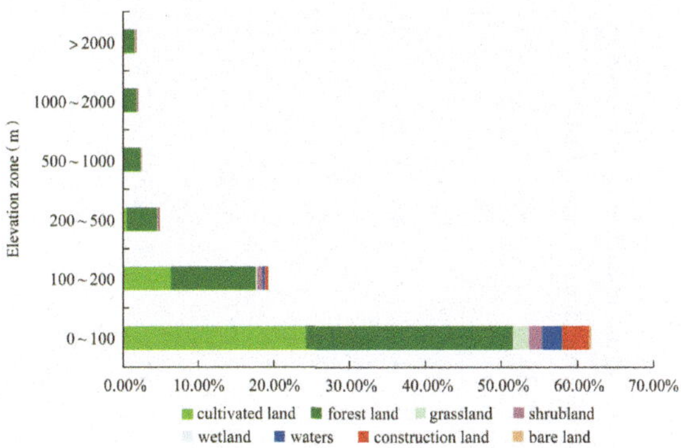

Fig. 17 Different land use by elevation zones

Table 5 Proportion of land use in each elevation zones (%)

Elevation zone (m)	Cropland	Forest land	Grassland	Shrubland	Wetland	Waters	Built-up land	Bare land	Total
<100	39.33	44.07	3.28	2.74	0.09	4.21	5.75	0.53	100.00
100–200	33.27	58.07	1.10	3.92	0.03	1.76	1.78	0.07	100.00
200–300	10.69	83.22	0.87	3.50	0.01	1.10	0.60	0.00	100.00
300–400	4.99	89.59	0.90	3.48	0.00	0.27	0.76	0.00	100.00
400–500	5.47	84.45	1.30	4.62	0.00	1.90	2.00	0.25	100.00
500–600	5.70	85.09	1.25	5.86	0.00	0.22	1.86	0.00	100.00
600–700	4.97	85.24	1.72	6.00	0.00	0.77	1.26	0.05	100.00
700–800	7.32	83.20	2.13	6.39	0.00	0.14	0.78	0.04	100.00
800–900	9.23	78.99	2.97	8.00	0.00	0.09	0.72	0.00	100.00
900–1,000	13.85	72.38	4.31	8.47	0.00	0.10	0.89	0.00	100.00
1000–1,100	18.09	65.93	5.19	8.50	0.00	0.71	1.55	0.03	100.00
1,100–1,200	23.00	58.82	5.53	8.58	0.00	1.69	2.36	0.01	100.00
1,200–1,300	29.87	52.24	6.34	7.83	0.00	0.11	3.61	0.01	100.00
1,400–1,500	43.87	40.61	6.61	6.19	0.00	0.04	2.66	0.01	100.00
1,500–1,600	47.27	39.13	6.57	4.96	0.00	0.02	2.04	0.01	100.00
1,600–1,700	36.46	48.48	8.17	4.81	0.00	0.01	2.06	0.01	100.00
1,700–1,800	25.03	60.99	7.68	4.20	0.00	0.04	2.06	0.01	100.00
1,800–1,900	14.00	73.91	7.10	2.94	0.00	0.04	2.00	0.02	100.00
1,900–2,000	8.65	78.33	6.53	1.98	0.01	0.92	3.55	0.04	100.00
>2000	7.05	77.99	8.52	1.77	0.01	0.39	4.19	0.07	100.00

Table 6 Proportion of different land use by slope gradient (%)

Slope zone	Cropland	Forest Land	Grassland	Shrubland	Wetland	Waters	Built-up Land	Bare land
0°–2°	46.59	25.72	38.38	24.75	61.89	69.31	39.05	63.63
2°–5°	41.55	30.85	34.08	31.91	34.09	23.09	45.16	29.56
5°–15°	9.23	24.03	17.40	26.79	3.88	7.03	14.45	6.32
15°–25°	1.91	12.91	6.41	10.99	0.08	0.45	1.12	0.41
25°–35°	0.58	5.18	2.52	3.96	0.04	0.07	0.18	0.06
35°–55°	0.14	1.26	1.11	1.48	0.01	0.04	0.04	0.02
>55°	0.00	0.05	0.10	0.13	0.00	0.01	0.00	0.01
Total	100.00	100.00	100.00	100.00	100.00	100.00	100.00	100.00

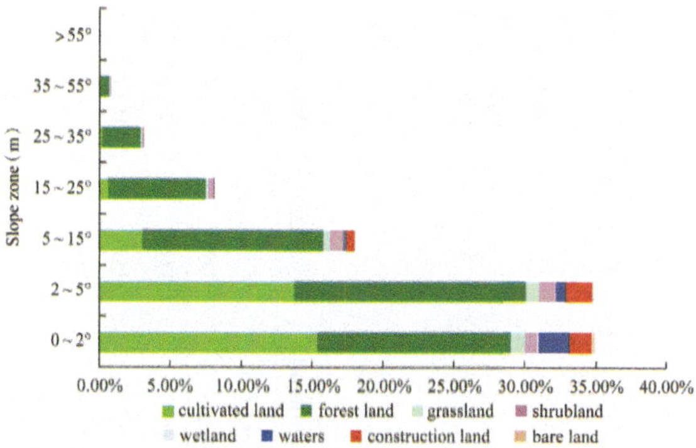

Fig. 18 Different land use by slope gradient

$$NDVI_{ij} = NDVI_i - NDVI_j$$

$NDVI_{ij}$ means the NDVI change value of j-th year compared to that of the i-th year. $NDVI_i$ and $NDVI_j$ are the NDVI value of i-th year and j-th year respectively.

- Regression Analysis

It is an important method to investigate the statistical relationship between multiple variables and to study the long-term change trend of vegetation. The following mathematical model can be used for a group of temporal independent variable x and NDVI dependent variable y.

$$y = a + kx_i + \varepsilon_i$$

a and k are unknown constants and ε_i is a random error. k can be calculated by $((x_i, y_i)(i = 1, 2, \ldots n))$, the unknown param.

Table 7 Proportion of land use in each slope gradient (%)

Slope zone	Cropland	Forest land	Grassland	Shrubland	Wetland	Waters	Built-up land	Bare land	Total
0°–2°	43.98	39.12	2.99	2.41	0.11	6.10	4.66	0.64	100.00
2°–5°	39.34	47.08	2.66	3.12	0.06	2.04	5.40	0.30	100.00
5°–15°	16.87	70.78	2.62	5.06	0.01	1.20	3.34	0.12	100.00
15°–25°	7.76	84.71	2.15	4.62	0.00	0.17	0.57	0.02	100.00
25°–35°	6.08	87.17	2.17	4.27	0.00	0.07	0.23	0.01	100.00
35°–55°	5.62	83.92	3.77	6.32	0.00	0.15	0.21	0.01	100.00
>55°	4.91	72.43	8.17	13.18	0.00	0.90	0.36	0.07	100.00

$$k = \frac{\sum_{i=1}^{n}(x_i - \bar{x})(y_i - \bar{y})}{\sum_{i=1}^{n}(x_i - \bar{x})^2}$$

Specifically, $\bar{x} = \frac{1}{n}\sum_{i=1}^{n} x_i$, $\bar{y} = \frac{1}{n}\sum_{i=1}^{n} y_i$.

The linear equation is obtained by linear fitting with the least square method of a set of NDVI observation values at the same pixel position. The slope of the equation,k shows the multi-year change trend of NDVI value of this pixel. If k is more than zero, vegetation increases; otherwise it decreases.

- Analysis on the slope (linear tendency rate) of annual average vegetation index change

The annual average vegetation index (NDVI) is the dependent variable and the year (t) is the independent variable. Thus the NDVI can be seen as a temporary function.

$$NDVI = f(t)$$

The function can be obtained through least square fitting. Assuming that the change is linear, the function can be expanded as follows:

$$NDVI = k * t + c,$$

The k stands for the slope and the c stands for the intercept.

If the year of 2000 is the initial year, the t equals 1. Given this, as one year goes by, t increases by 1, up to 19. NDVI takes the corresponding annual average vegetation index. The slope established by pixels can be used for analysis on the spatial change of vegetation. If the slope is less than 0, the pixel shows a negative trend of change with the pessimistic vegetation index, reflecting that the vegetation activity is restrained, and vice versa.

2.1 Spatial Distribution Characteristics of Vegetation Cover

The spatial distribution data of annual average vegetation index in Sri Lanka can be obtained based on its average level in a period of recent 19 years from 2000 to 2018 (Fig. 19).

Distribution Characteristics of Vegetation Cover

The annual average NDVI distribution image shows that the vegetation cover across the country is relatively high, with the highest in the central, southwestern and northwestern regions. The annual average NDVI of the country is 0.701, but the figure varies with spatial distribution. Specifically, the non-vegetation areas with NDVI less than 0.400 are mainly those at the coastal regions: saline alkali land, inland surface water and built-up land. The low vegetation-cover areas with NDVI between 0.400

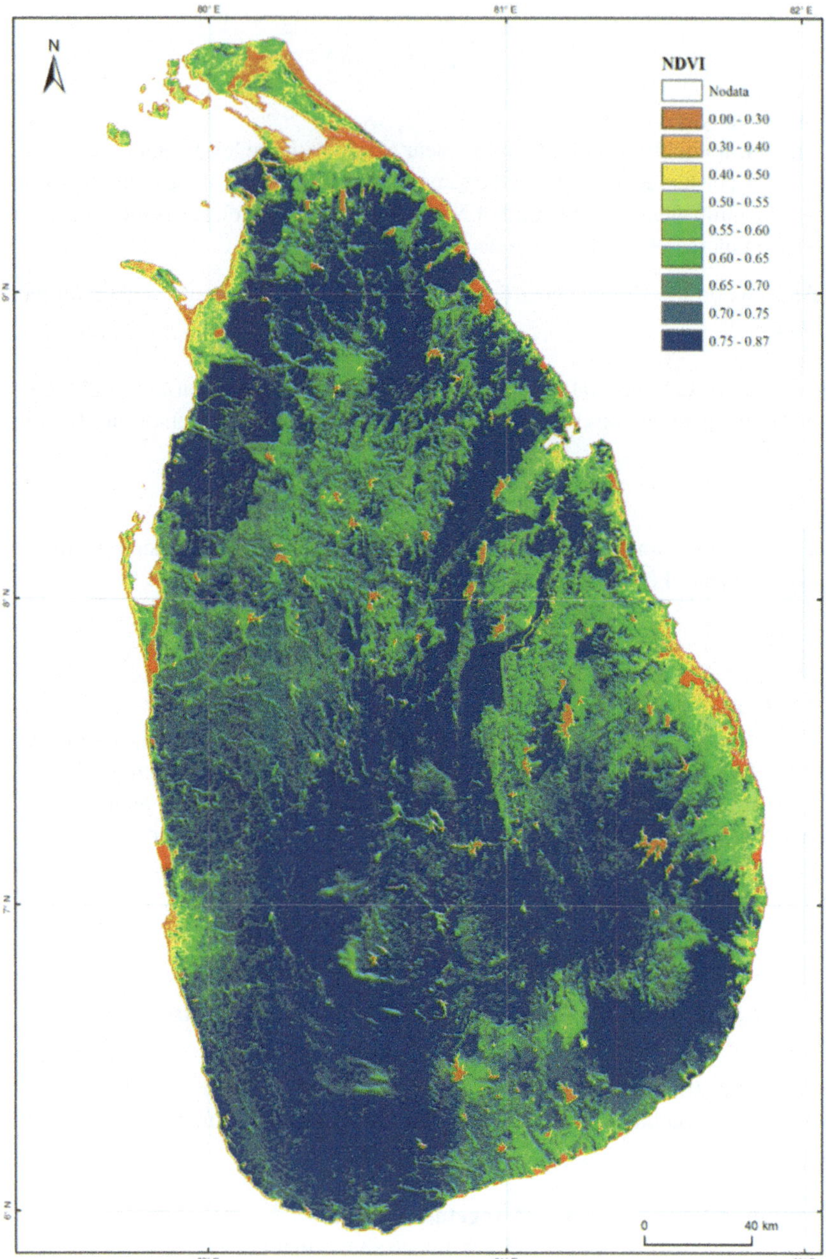

Fig. 19 Distribution image of annual average NDVI

and 0.550 are grasslands, sparse shrubs, abandoned land and built-up land with little vegetation. Moreover, the cultivated areas mainly represent the moderate vegetation-cover areas with NDVI ranging between 0.550 and 0.700, and forest land make up the high vegetation-cover areas with NDVI values greater than 0.700 (Fig. 20).

Five of the nine provinces have annual average NDVI higher than that of the whole country, all located in the southwest monsoon region, have Sabaragamuwa with the highest index. Among the other four provinces having NDVI index below the national average, Northern and Eastern provinces in the arid regions have the lowest (Fig. 21).

Distribution Characteristics of Vegetation Cover by Provinces

Among the nine provinces, Northern and Eastern provinces have higher proportion of non-vegetation areas of 10.43% and 4.96% respectively mainly lagoons and saline alkali lands along their coasts. Also, Western, North Western and Southern provinces have large non vegetated areas mainly water bodies and built-up in Western and Southern provinces, and coastal beaches and lagoons in North Western Province. Conversely, the proportion of the non-vegetation areas in the other provinces is relative low, mainly rivers, lakes, reservoirs and ponds.

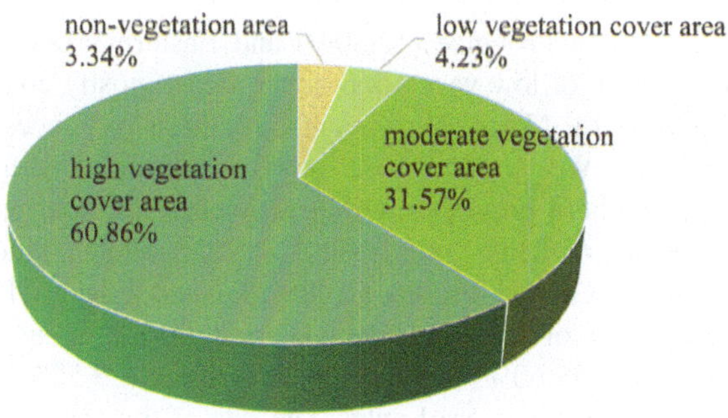

Fig. 20 Proportion of the annual average NDVI area

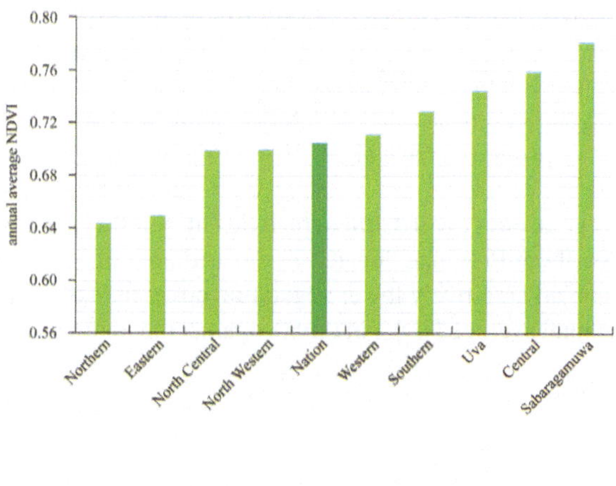

Fig. 21 Comparison of annual average NDVI between Sri Lanka and provinces

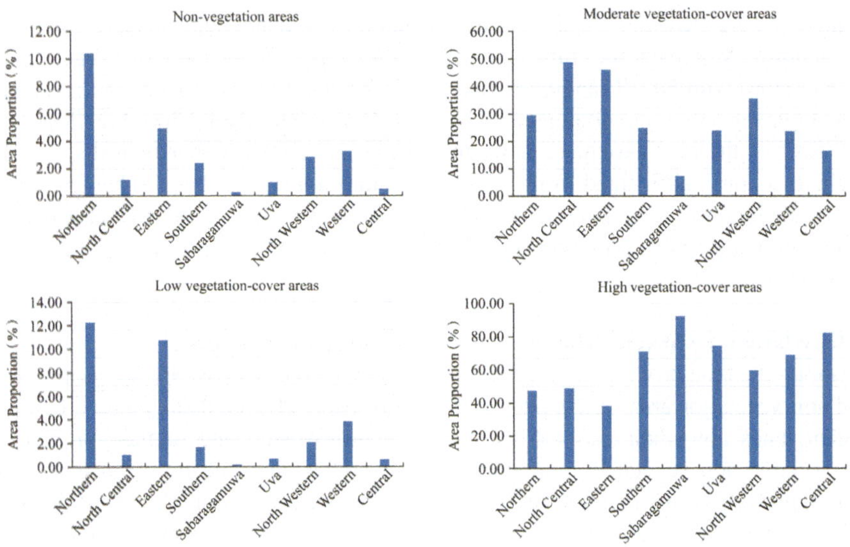

Fig. 22 Proportion of vegetation cover areas by provinces

Northern Province (12.30%) and Eastern Province (10.78%) have the highest proportion of low vegetation-cover areas, mostly coastal salinized lands. Followed by Western, North Western and Southern provinces with salinized coastal lands in North Western Province, and built-up land with little vegetation in Western and Southern provinces. Other provinces have extremely low proportions, mostly lakeside grasslands, sparse shrubs and abandoned lands.

North Central Province (48.68%) and Eastern Province (46.03%) have high proportion of the moderate vegetation-cover areas, followed by North Western Province (35.60%) and Northern Province (29.57%). These are the major agricultural provinces in Sri Lanka. The rest of the provinces, except for Sabaragamuwa and Central provinces (<20%), also account for more than 20%.

With generally high proportion of high vegetation-cover, five of the nine provinces have higher proportions than the national average. The high vegetation cover dominates each provincial land except Eastern Province which is dominated by moderate vegetation-cover (46.03%). Sabaragamuwa Province (92.18%) ranks first while the other provinces have about 50% or more (Figs. 22 and 23).

Vegetation Cover Characteristics of Northern Province

Northern Province, located in the arid area of Sri Lanka has more coastal salinized land and relatively lower vegetation cover than the national average, with the overall annual average NDVI of 0.650, against the national average of 0.701. In addition, the proportions of non-vegetation cover areas and low vegetation cover areas are significantly higher than the national average, while that of moderate and high vegetation cover areas are lower than the national average (Figs. 24 and 25).

Vegetation Cover Characteristics of North Central Province
North Central Province, located in the semi-arid area, has irrigated fields and forest lands, and its overall annual NDVI (0.696) is closer to that of national average (0.701). In addition, the proportions of non-vegetation and low vegetation-cover areas are significantly lower than the national average, while the moderate vegetation cover areas dominated by crops is 1.5 times the national average. The high vegetation-cover areas equals four fifth of the national average (Figs. 26 and 27).

Vegetation Cover Characteristics of Eastern Province
Eastern Province, located in the eastern costal region, has more salinized land, and the annual average NDVI (0.655) is lower than the corresponding national average. In addition, the proportions of non-vegetation and low vegetation-cover areas are significantly higher than the national average, while the moderate vegetation cover areas is nearly 1.5 times the national average. The high vegetation-cover areas is only 63% of the national average (Fig. 28).

Vegetation Cover Characteristics of Southern Province
Southern Province, located in the southern coastal area, has a higher annual average NDVI (0.721) than the national average. Geographically, its eastern region falling in the arid area has low vegetation cover while its western region located in the southwest monsoon area has abundant rainfall, lush forest land and high vegetation cover. Moreover, the proportions of non-vegetation cover, low vegetation-cover and moderate vegetation cover areas are much lower than the national average, while that of high vegetation-cover areas is higher (Figs. 29 and 30).

Fig. 23 Comparison of the proportion of different vegetation cover between Northern Province and the country

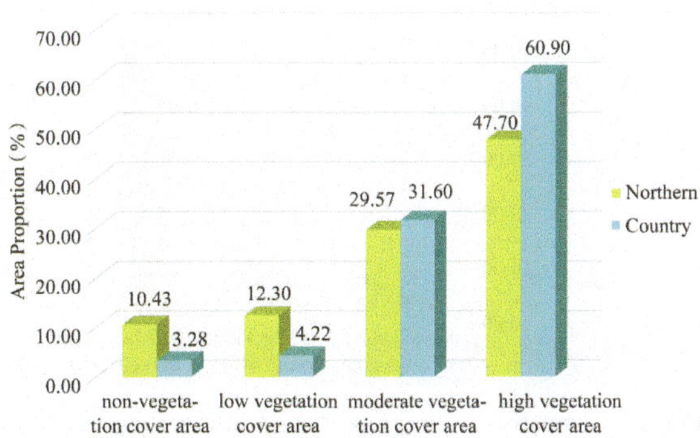

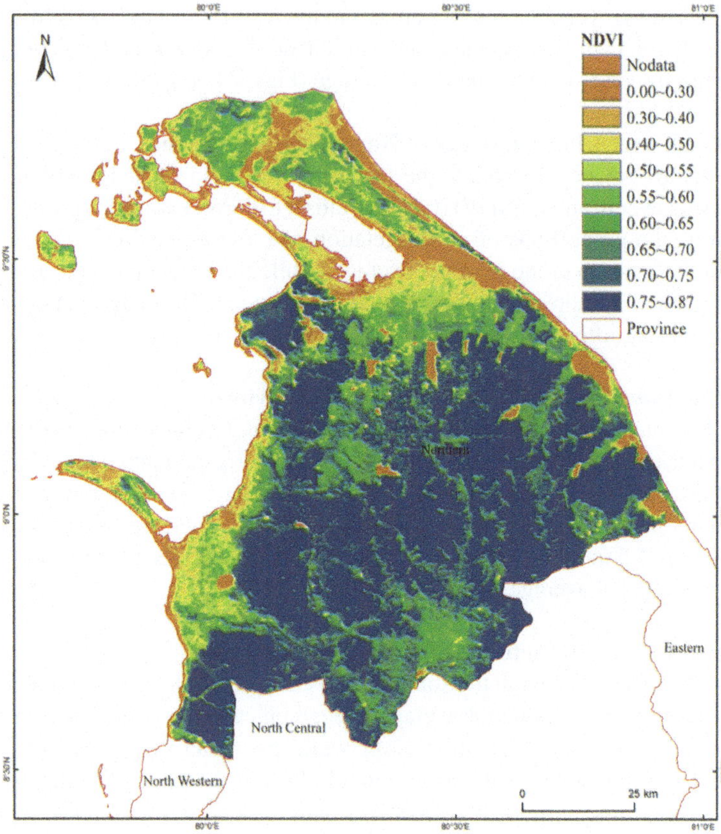

Fig. 24 Distribution image of annual average NDVI in Northern Province

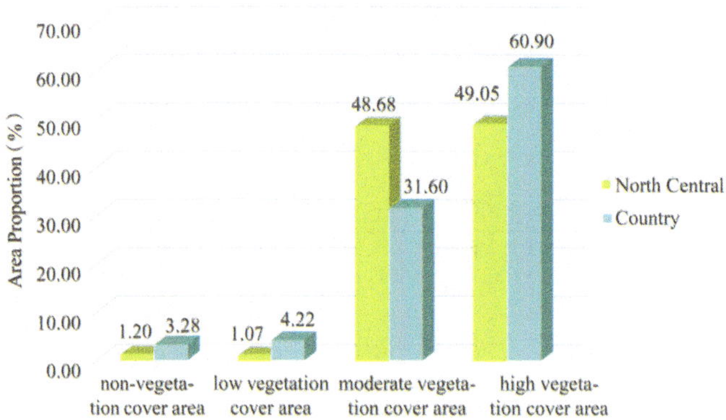

Fig. 25 Comparison of the proportion of different vegetation cover between North Central Province and the country

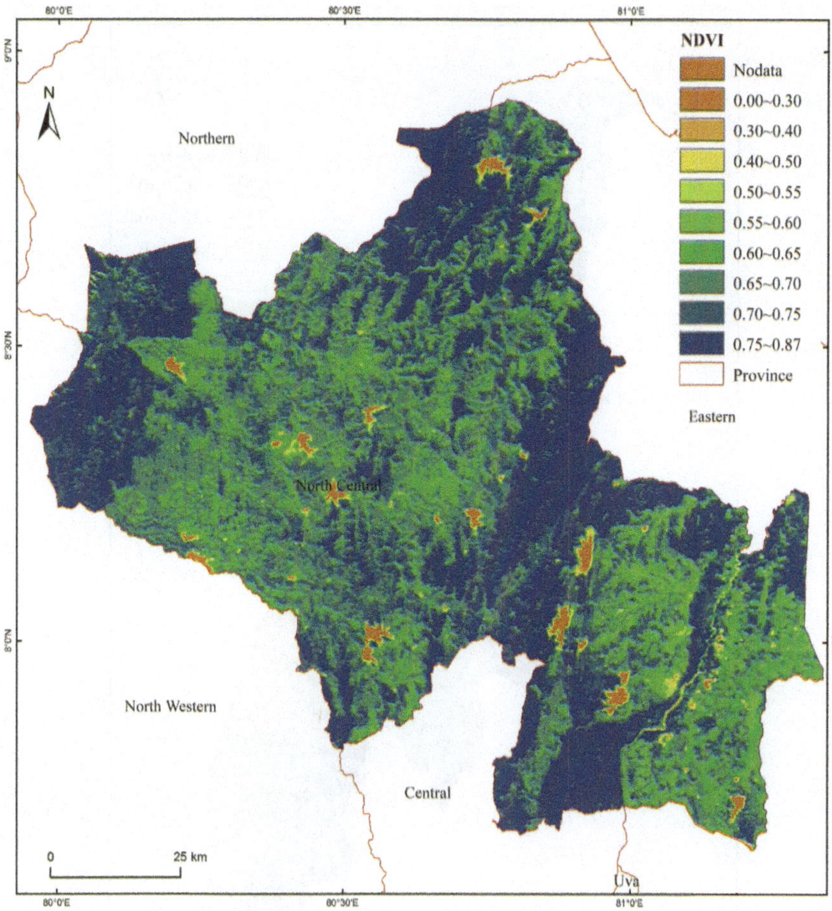

Fig. 26 Distribution image of annual average NDVI in North Central Province

Vegetation Cover Characteristics of Sabaragamuwa Province

Sabaragamuwa Province, located in the southwestern monsoon area, has abundant rainfall, lush forest and high vegetation cover, with its yearly average NDVI of 0.765 much higher than the national average. Also, the proportions of non-vegetation cover low vegetation-cover and medium vegetation-cover areas are much lower than the national average, while the high vegetation-cover areas is more than 1.5 times that of the national average (Figs. 31 and 32).

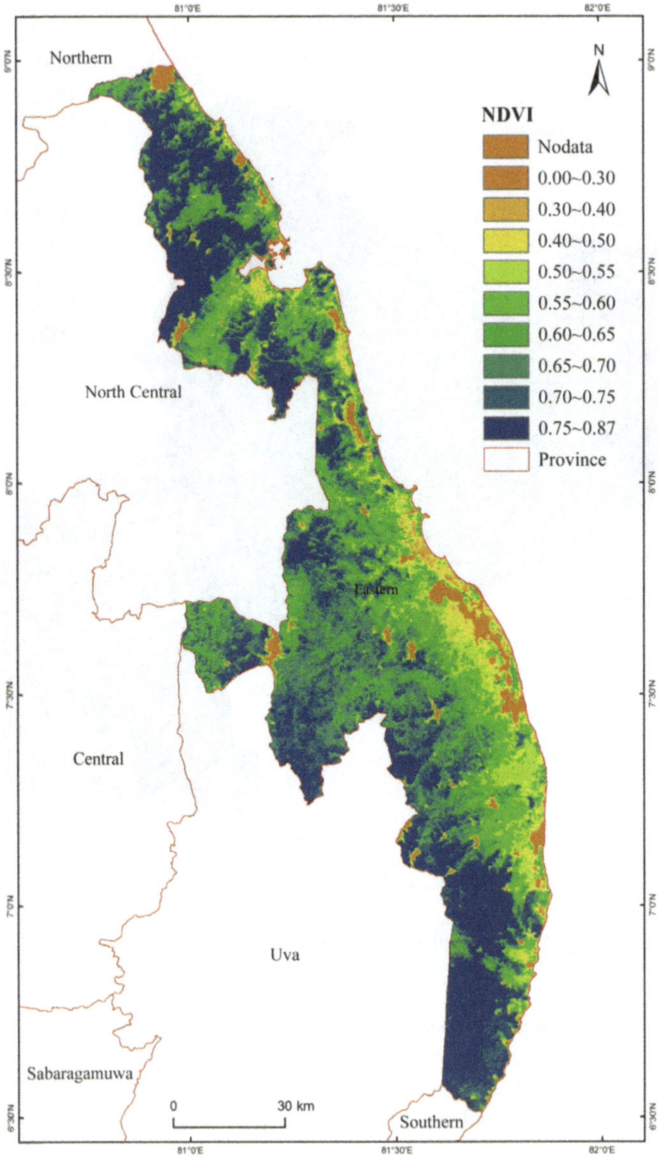

Fig. 27 Distribution image of annual average NDVI in Eastern Province

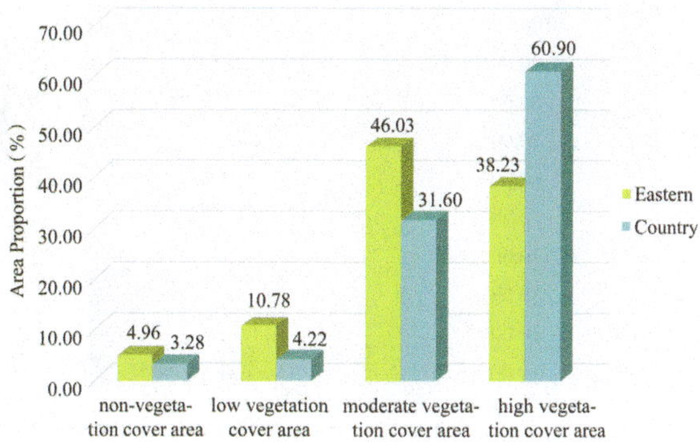

Fig. 28 Comparison of the proportion of different vegetation cover between Eastern Province and the country

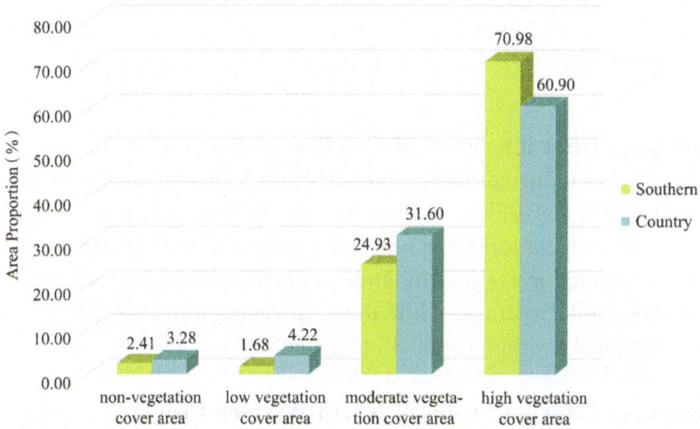

Fig. 29 Comparison of the proportion of different vegetation cover between Southern Province and the country

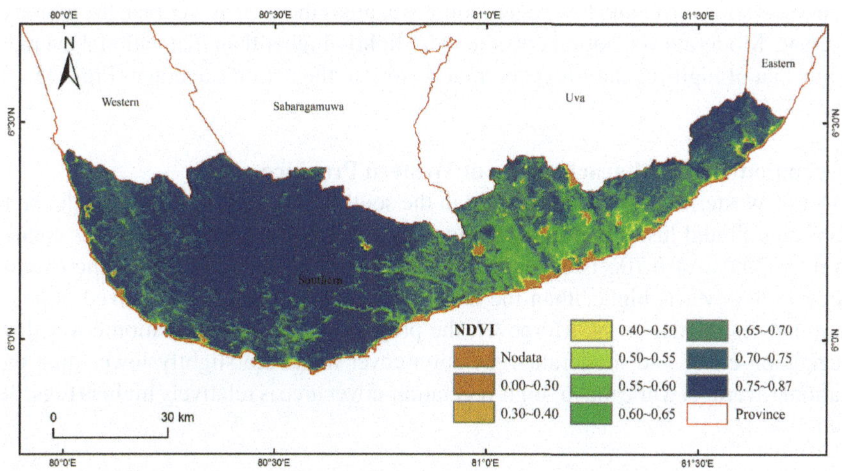

Fig. 30 Distribution image of annual average NDVI in Southern Province

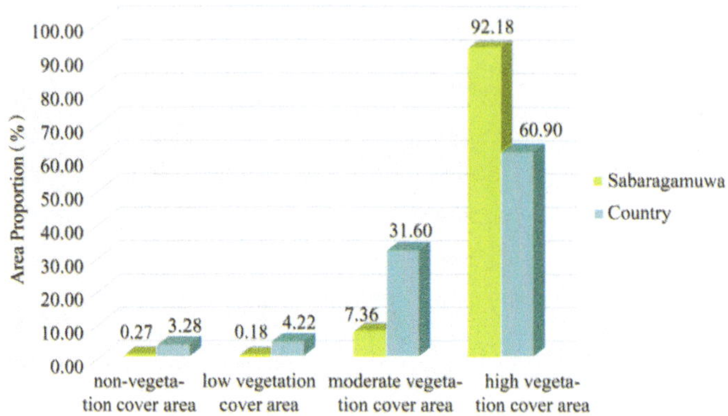

Fig. 31 Comparison of the proportion of different vegetation cover between Sabaragamuwa Province and the country

Vegetation Cover Characteristics of Uva Province

Uva Province houses luxuriant forest and fairly high vegetation cover, with its overall annual average NDVI of 0.734, higher than the national average. In addition, the proportions of non-vegetation cover and low vegetation-cover areas are significantly lower than the national average. Moderate vegetation cover areas (23.97%) is also lower than that national average, while the high vegetation-cover areas is more than 1.2 times the national average (Figs. 33 and 34).

Vegetation Cover Characteristics of North Western Province

North Western Province has the yearly average NDVI of 0.697 which is lower than the national average, with half of the province located in the semi-arid area mostly for irrigation field and with relatively few forest lands. Additionally, the proportions of non-vegetation cover and low vegetation-cover areas there are lower than the national average. Moderate vegetation cover area is slightly higher than that national average while that of high vegetation-cover area is close to the nation's average (Figs. 35, 36 and 37).

Vegetation Cover Characteristics of Western Province

Most of Western Province are located in the southwest monsoon region, with abundant rainfall and lush vegetation. Although it has large residential areas, its annual average NDVI of 0.706 is equivalent to that of the whole country and the overall vegetation cover is higher than the national average due to its centralized population and residential areas. Moreover, the proportions of non-vegetation cover, low vegetation-cover and moderate vegetation-cover areas are slightly lower than the national average, while that of high vegetation-cover area is relatively higher (Figs. 38 and 39).

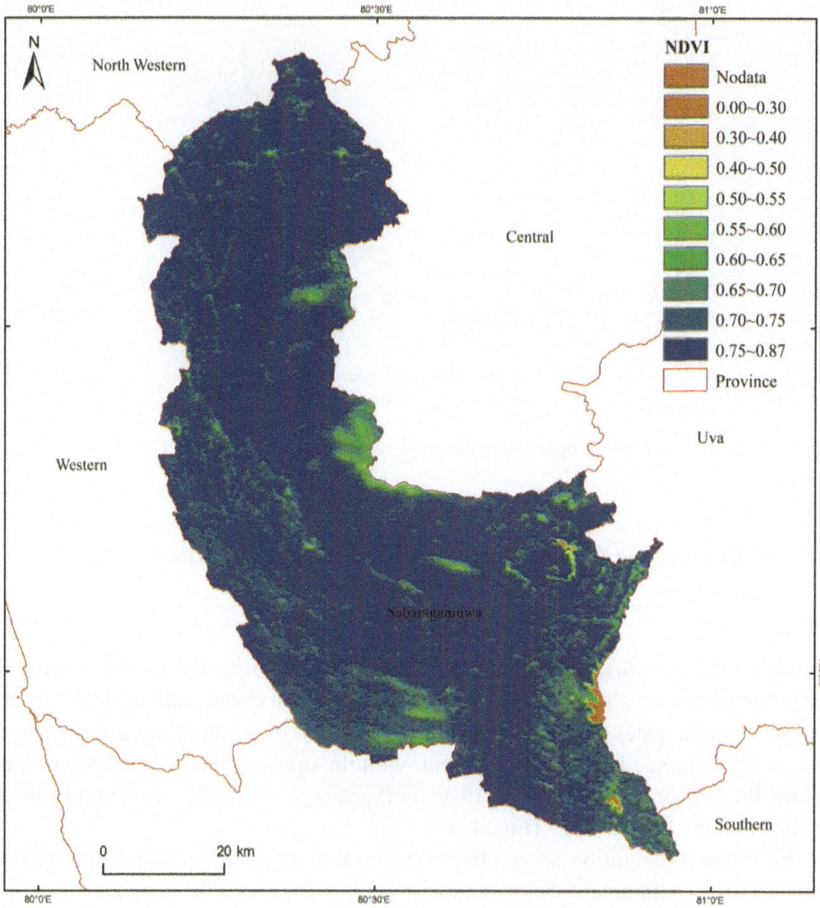

Fig. 32 Distribution image of annual average NDVI in Sabaragamuwa Province

Vegetation Cover Characteristics of Central

Central Province, located in the Central Highlands with lush forests, has much higher annual average NDVI of 0.746 than the national average. Additionally, the proportions of non-vegetation cover and low vegetation-cover areas are a little lower than the national average while that of moderate vegetation cover area is just half of the national average and that of high vegetation-cover area is about 1.35 times the nation's average (Fig. 40).

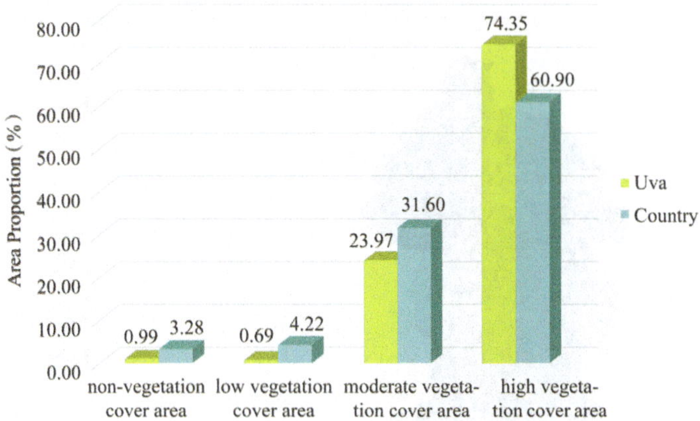

Fig. 33 Comparison of the proportion of different vegetation cover between Uva Province and the country

2.2 *Quantitative Characteristics of Vegetation Cover Variations*

In light of the quantitative characteristics of annual change, the overall vegetation cover has increased slightly since 2000, with a linear increase rate of 0.01 for each decade. That indicates that the vegetation cover across the country has become better. The trend of change by years is different, with the upward trend from 2000 to 2004 and the fluctuation from 2004 to 2018. Over the past 19 years, the worst trend appears in 2000 and the best in 2015 (Fig. 41).

The national vegetation cover, affected by rainfall, crop maturity and other factors, shows obvious characteristics of annual monthly change. The vegetation cover is relatively low from May to September. Periodically, monthly NDVI in August is the lowest (0.645), and begins to increase from October and reaches the highest in January (0.768) and then a gradual decrease (Fig. 42).

2.3 *Spatial-Temporal Dynamics of Vegetation Cover*

The overall quantitative analysis can reflect the characteristics of vegetation change in a certain region, but the long-term change trend, change quantity and change intensity of vegetation by regions are different in space and time. Figure 43 shows the annual average NDVI distribution image from 2000 to 2018, which directly reflect the spatial distribution of yearly vegetation cover in Sri Lanka. In order to analyze the spatial-temporal dynamics of vegetation cover in Sri Lanka, this section, based

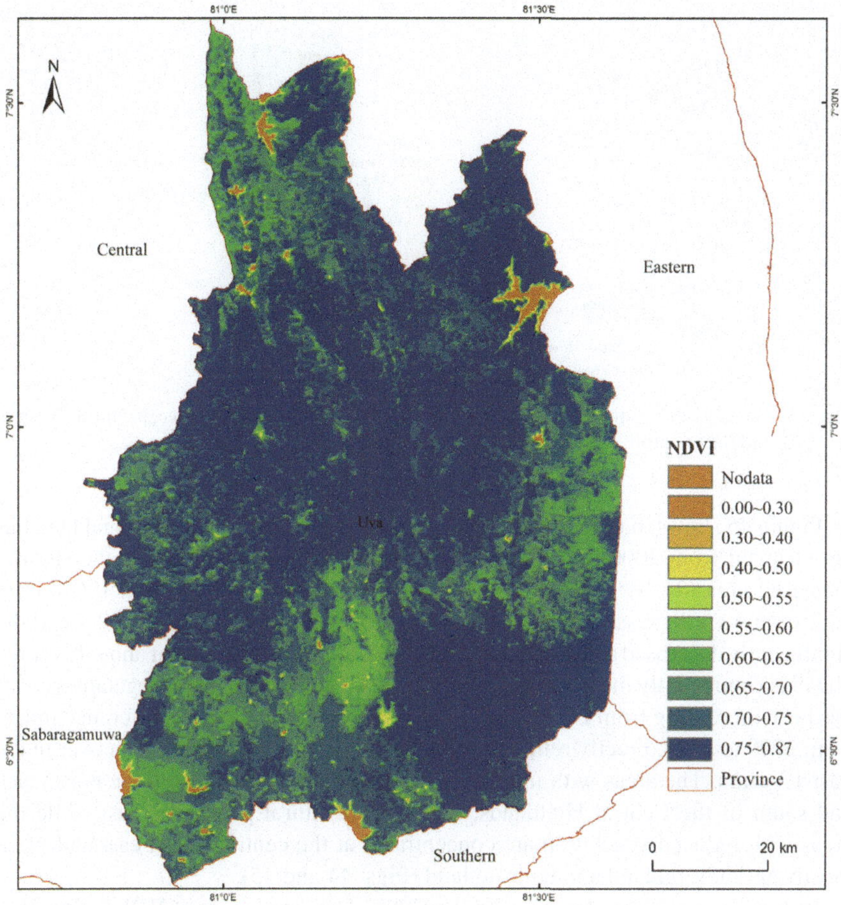

Fig. 34 Distribution image of annual average NDVI in Uva Province

on the mathematical statistics and the spatial analysis of GIS, discusses its spatial-temporal dynamics and differences from the perspectives of image difference, linear change trend, and annual absolute rate of change (Fig. 43).

Difference Analysis of Annual Average NDVI from 2000 to 2018

According to the time span of MOD13Q1-NDVI data, the years of 2000, 2005, 2010, 2015 and 2018 are selected as the five-year interval of differentiated analysis. On the basis of the five-year average NDVI data, four sub periods of 2000–2005, 2005–2010, 2010–2015, 2015–2018 and the whole period of 2000–2018 can be obtained. From the value range, vegetation change can be divided into five grades: obvious decrease (NDVI < −0.100), slight decrease (−0.100 ~ −0.050), basically stable (−0.050 ~ 0.050), slight increase (0.050–0.100) and obvious increase (NDVI > 0.100).

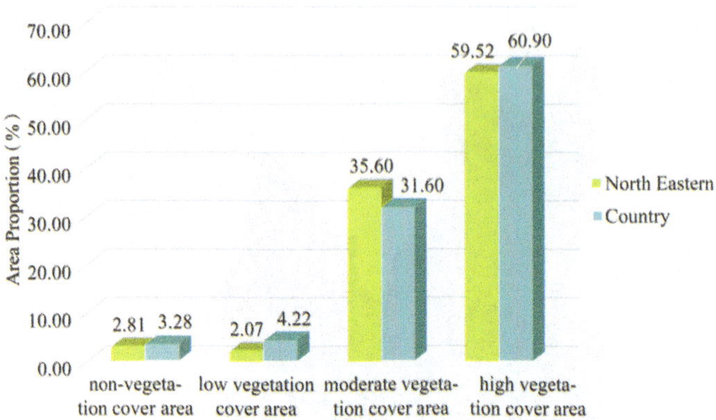

Fig. 35 Comparison of the proportion of different vegetation cover between North Western Province and the country

Figure 45 shows that the annual average NDVI of 49.59% of the national land has not changed from 2000 to 2018; 45.20% shows an increase, of which the slightly increased area (31.10%) is twice as more as the obvious increase (14.10%); only 5.22% shows a decrease, of which the slightly decreased area covers 4.02% and the significantly decreased area is as small as 1.20%. The increased area of annual average NDVI is significantly more than the reduced area, indicating that the vegetation cover has been increasing from 2000 to 2018. The annual average NDVI difference image from 2000 to 2018 directly reflects the spatial distribution of vegetation cover in the past 19 years. The areas with increased vegetation are distributed in the north, east and south of the Central Highlands, mostly agricultural planting areas, while the areas with reduced vegetation are concentrated at the central north, east and west, mostly agricultural land and built-up land (Figs. 44 and 45).

During the first five years from 2000 to 2005, the annual average NDVI of 58.75% of Sri Lanka's area remained unchanged; 38.21% showed an increase, of which the area with slight increase accounted for 30.16%, while the area with obvious increase was relatively small (8.05%). The area with slight increase and obvious increase dominated the four periods. 3.04% of the area showed a decline, where the area of slight reduction was 2.48% and the area of obvious reduction was the least, covering 0.56%. The proportion of area of slight reduction and obvious reduction was the least in four periods. It may be concluded that the area of national annual average NDVI increase is far greater than the area of decrease. The average NDVI difference from 2000 to 2005 reflect the increased area of annual NDVI as opposed to the decreasing areas. The increase area is mainly in the north central and southwestern regions, particularly in the middle reaches of Mahaweli River, while the decreased areas are scattered in the low-altitude agricultural planting areas.

The annual average NDVI area showed stability during the second five-year period from 2005 to 2010, which was the period with the largest proportion of stable area (78.91%). Although significantly less than the previous five-year period, 9.94% of the

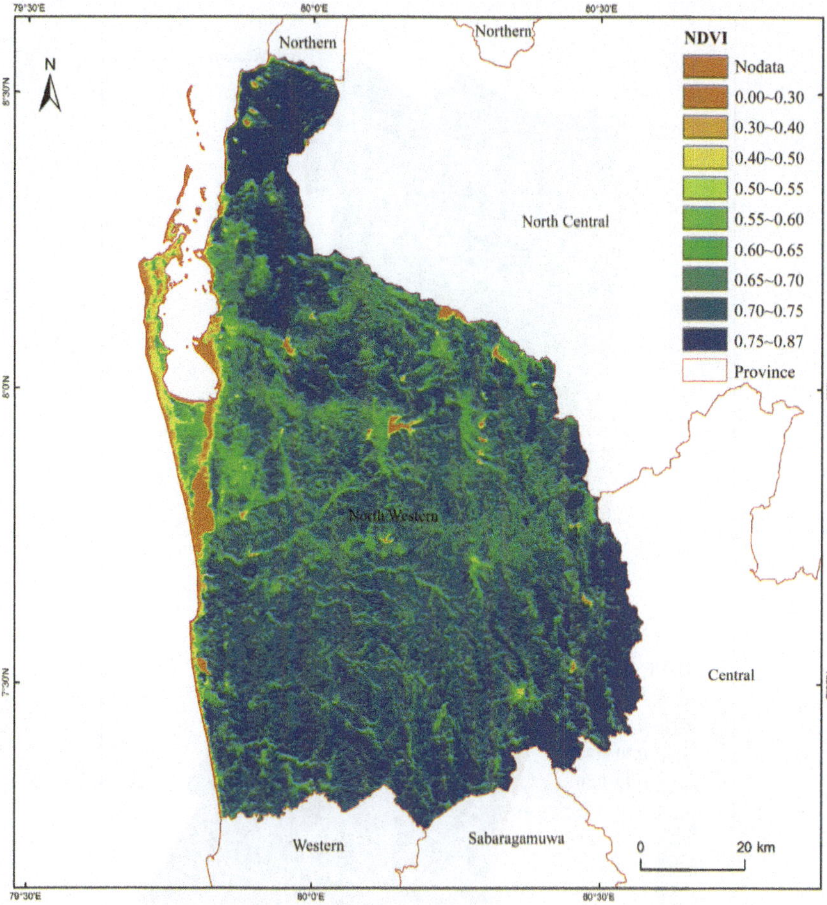

Fig. 36 Distribution image of annual average NDVI in North Western Province

area showed an increase, in which the area with more slight increase area (8.69%) than significant increase area (1.26%). Moreover, larger than that of the previous five-year period, 11.15% showed a decline, of which the area of slight decrease was 9.15% but the area of obvious decrease was only 2.00%. Notably, the area of decreased national annual average NDVI is slightly more than that of increased areas, with areas of increase mainly distributed in the middle west, the middle east and the north, while the areas of decrease are mostly the middle north, the southwest and the Central Highlands.

In the third five-year period from 2010 to 2015, the stable area of annual average NDVI is 74.46%, although slightly less than the preceding five-year period. 19.75% of the area showed an increase, where the area with slightly increase was16.31% and the area with significant increase was 3.45%. Further, 5.78% of the area showed a decrease, where the area with slightly decrease was 4.70% and that with obvious

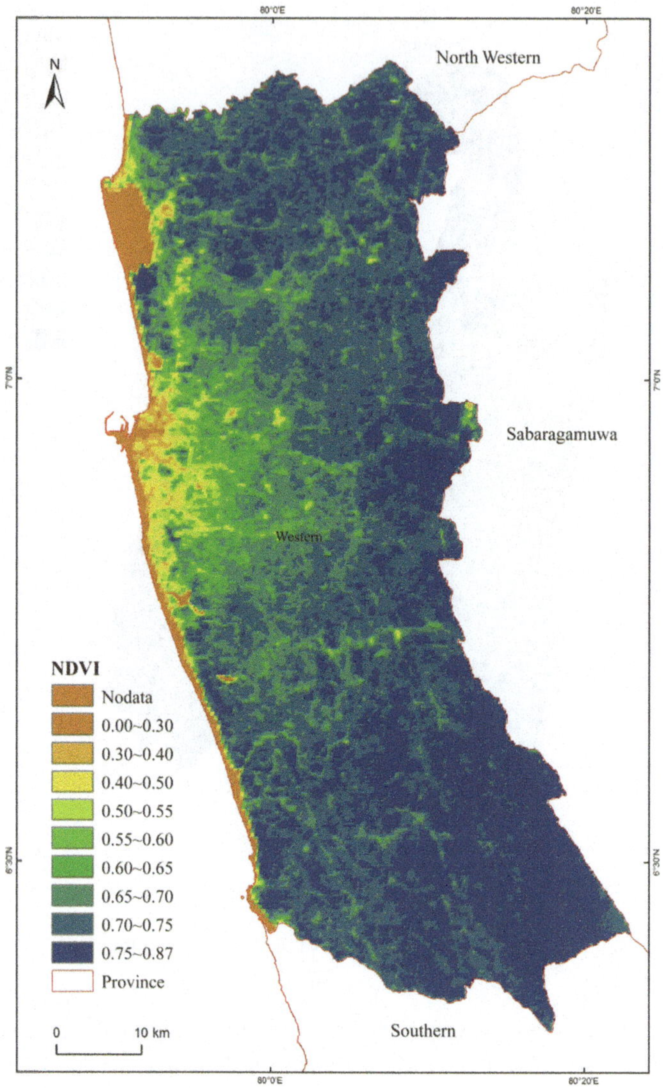

Fig. 37 Distribution image of annual average NDVI in Western Province

decrease was the least, accounting for 1.08%. Generally, the area with increased annual average area of NDVI was significantly more than that with decreased value in this period as depicted by 2010–2015 average NDVI value difference. The increased areas were mainly distributed in the northwest, west, southwest and southeast, among which the northwest coastal area had an obvious increase. On the other hand, the areas of decrease were mainly distributed in the middle east and middle north.

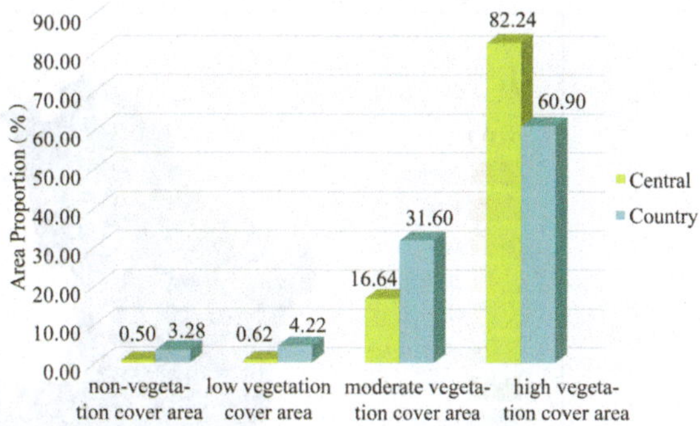

Fig. 38 Comparison of the proportion of different vegetation cover between Western Province and the country

In the fourth period from 2015 to 2018, the annual average NDVI area was basically stable at 77.71%, almost similar to that of the second five-year period. Although significantly less than that of the previous five-year period, 5.91% of the area showed an increase and is the period with the least proportion of both slight and significant increase. 16.37% showed a decrease and thus the period with the highest proportion of significant and slight decrease. Notably, the area with decreased average NDVI is much more than the corresponding increase with annual average NDVI reduction mainly in the coastal areas of northwest, the west and southwest, while increase zones are mainly distributed in the agricultural planting areas.

The variation characteristics of annual average NDVI differ by periods. In the first five-year period, the area with increase value was 11.58 times larger than the decrease area, hence overall vegetation cover increase. In the second period, the increased-NDVI area is slightly smaller than the decreased one, hence little overall vegetation cover change. In the next five-year period, the increased annual average areas was 3.42 times the decrease area, although less than the proportion of the first period. For final period, the increased-NDVI area was far less than that in the decreased one and the vegetation cover became worse. Their common characteristics is that the area proportion of the unchanged NDVI ranks first.

Analysis on the Linear Trend of Annual Average NDVI from 2000 to 2018

To reflect the continuous and spatio-temporal change trend of NDVI, the annual average NDVI time series data from 2000 to 2018 is used for linear fitting of each pixel through least square regression analysis, and then the slope diagram of each pixel fitting line is calculated. Compared with the image difference method, it is more accurate to analyze the trend of long-term change of time series data as the least square regression analysis based on time series data can effectively reduce the huge error influence caused by the uncertain factors. Considering the deviation of MOD13Q1 data and the effectiveness of pixel representation for vegetation free land, we define the slope of linear change as follows in order to facilitate the analysis of linear change trend of vegetation (Table 8).

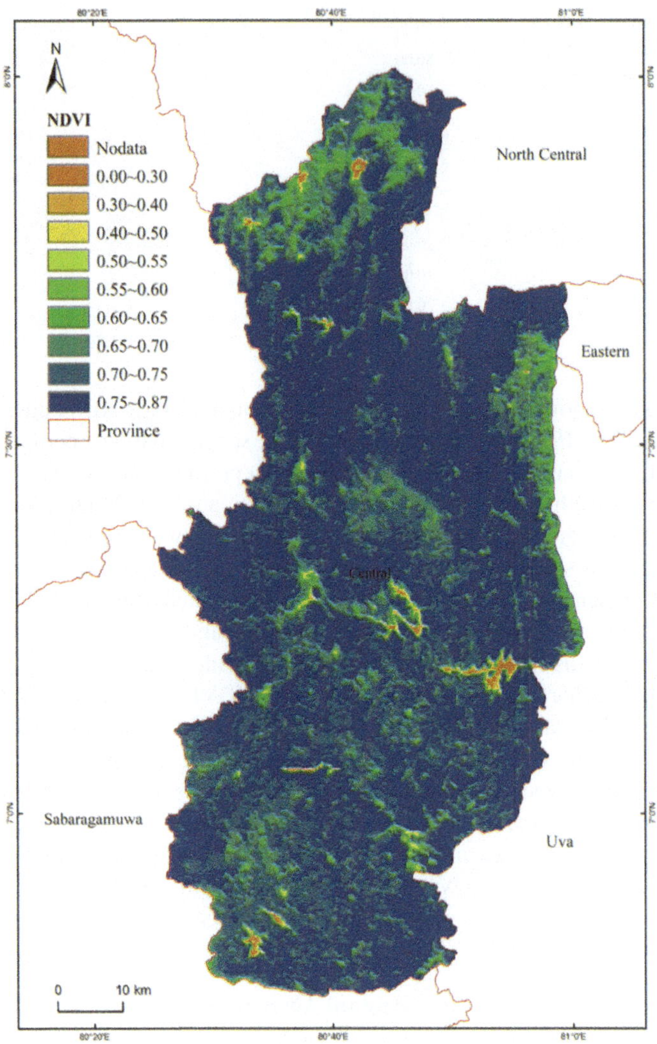

Fig. 39 Distribution image of annual average NDVI in Central Province

The linear change trend chart of vegetation from 2000 to 2018 and the area proportion of each level can be generated from above definition. The analysis chart outlines that 53.27% of the national land has an increasing NDVI, of which 58.49% of the pixel change slope calculated through $a = 0.1$ shows a significant increase, while the remaining failing to pass significance level test indicates that the vegetation change fluctuates frequently and its linear increasing trend is not obvious. Contrastingly, 16.07% shows a decreasing trend, where 43.88% of the pixel change slope obtained

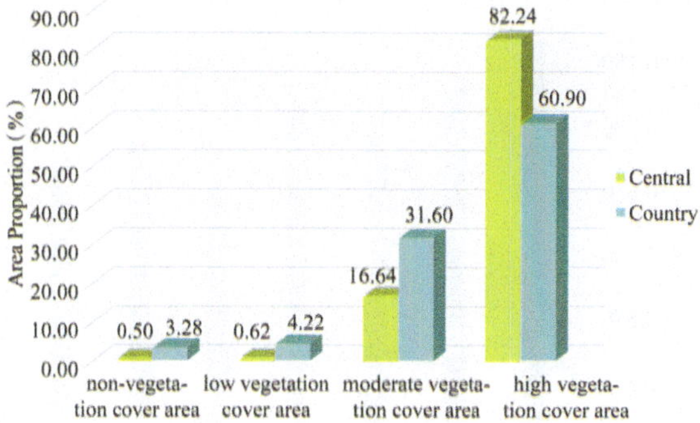

Fig. 40 Comparison of the proportion of different vegetation cover between Central Province and the country

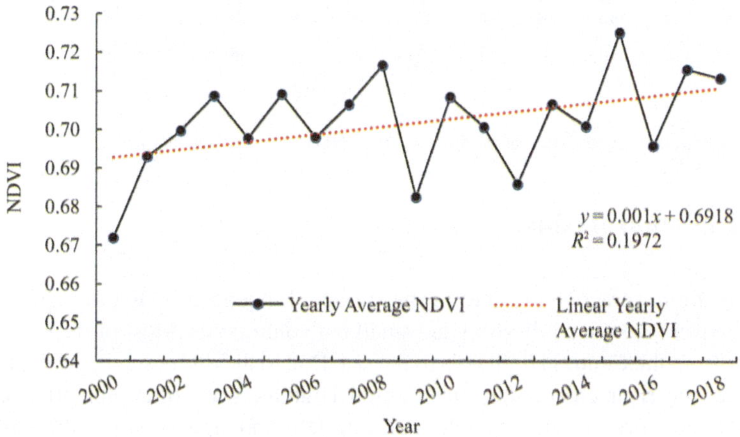

Fig. 41 Change trend of annual average NDVI

through $a = 0.1$ shows a significant reduction, while 56.12% failing to pass significance level test indicates that frequent fluctuations occur in vegetation change and its linear decrease trend is not obvious. Additionally, 30.66% of the area has the figure between −0.001 and 0.001, meaning that its vegetation remains unchanged. The areas with increasing vegetation cover are mainly located in the west of Uva, the east of Sabaragamuwa, the northeast of Central, the south of North Central and the west of East provinces. Conversely, the areas with decreasing vegetation cover are mostly agricultural planting area, built-up land and its surrounding areas, including the eastern coastal areas of Western, Northern, North Central, North Western and East as well as the southwest of Central provinces (Fig. 46).

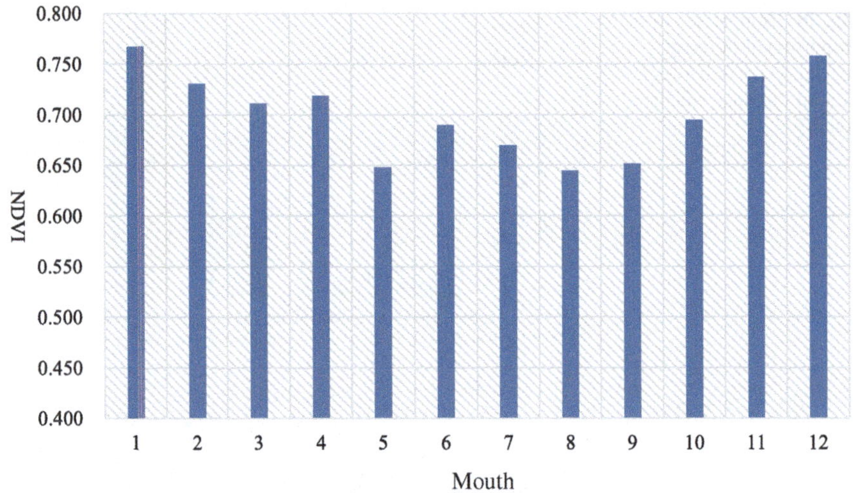

Fig. 42 Change characteristics of monthly average NDVI

3 Rivers and Basins

Water is the most active element in the surface. Influenced by the topography, Sri Lanka's rivers are relatively short and small but voluminous. Most of the rivers flow radially from the Central Highlands to the sea. Due to the discontinuity of geological structure, the river channels in the Central Highlands are often cut off from the main channel, forming a large number of waterfalls and rapids on the cliffs and steep slopes. As they flow through floodplains and deltas, the velocity gradually decreases. Despite being short and small, they all are important for irrigation. The longest river is Mahaweli River, with a total length of 335 km (Fig. 47).

Sri Lanka has 103 major river basins, accounting for 90% of the total land. However, the remaining 10% are located along the coast and Jaffna Peninsula. Most of the river basins are small watersheds, only 17 of the 103 watersheds cover an area of more than 1000 km^2 and 11 cover an area of 500–1000 km^2. More importantly, Mahaweli River, the longest river in Sri Lanka, has the largest drainage area of 10,448 km^2 (Fig. 48 and Table 9).

Source: Food and Agriculture Organization of the United Nations.

There are 16 rivers with a length of more than 100 km in the country, mostly originated from the Central Highlands (Fig. 49 and Table 10).

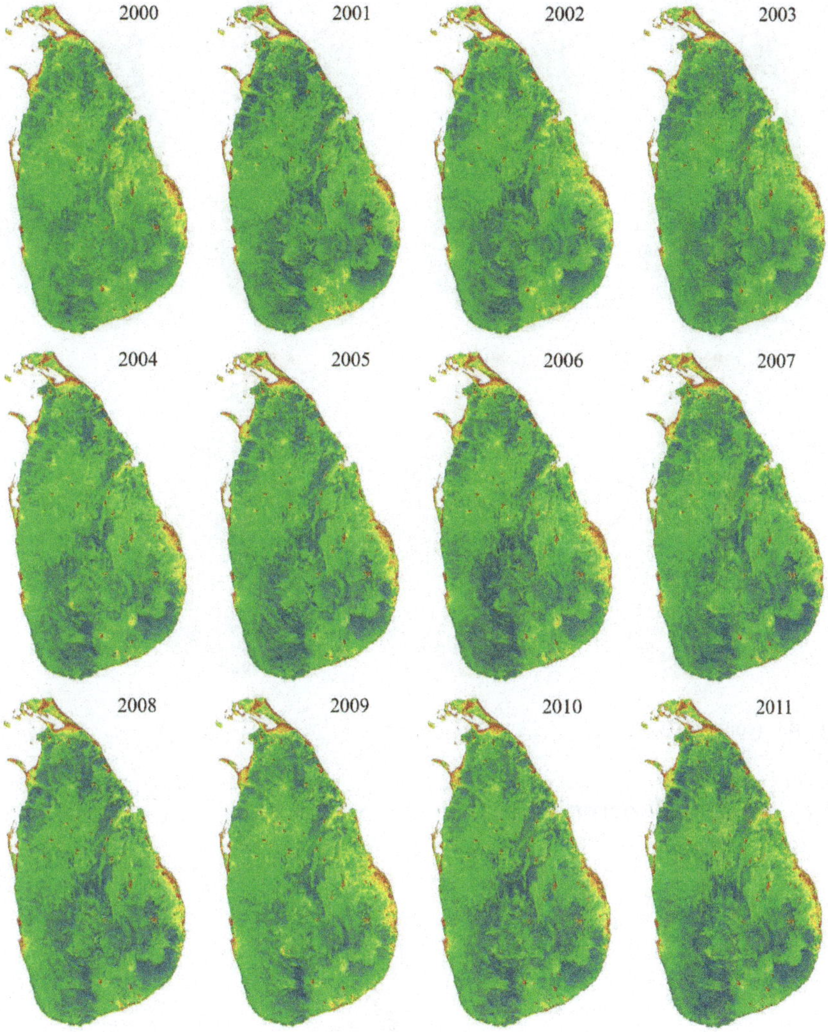

Fig. 43 Distribution image of annual average NDVI

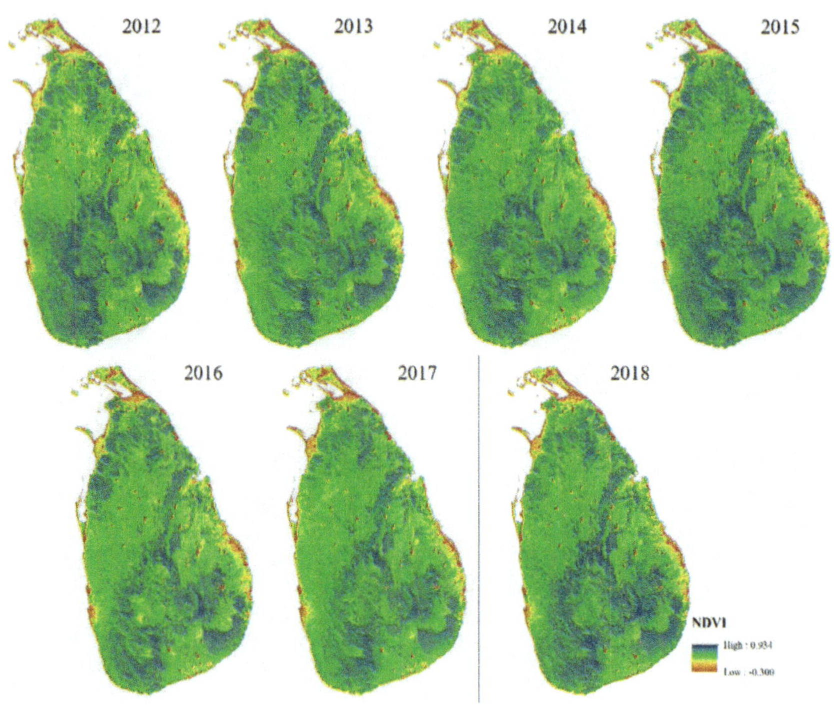

Fig. 43 (continued)

3.1 *Mahaweli River*

Mahaweli River (Sinhalese: "Great Sandy River") is the longest river nationwide, accounting for about one sixth of the land area. The river, sourcing from Horton Plains, flows to the north, and turns to the east near Kandy; it then turns to the northeast around Handaganawa, and finally enters the sea near Koddiyar Bay, along which there are mountains, valleys, waterfalls and plains (Figs. 50 and 51).

Mahaweli River is located in the middle east of Sri Lanka between 6° 46′ N–8° 28′ N and 80° 28′ E–81° 20′ E. The terrain of the basin is high in the South

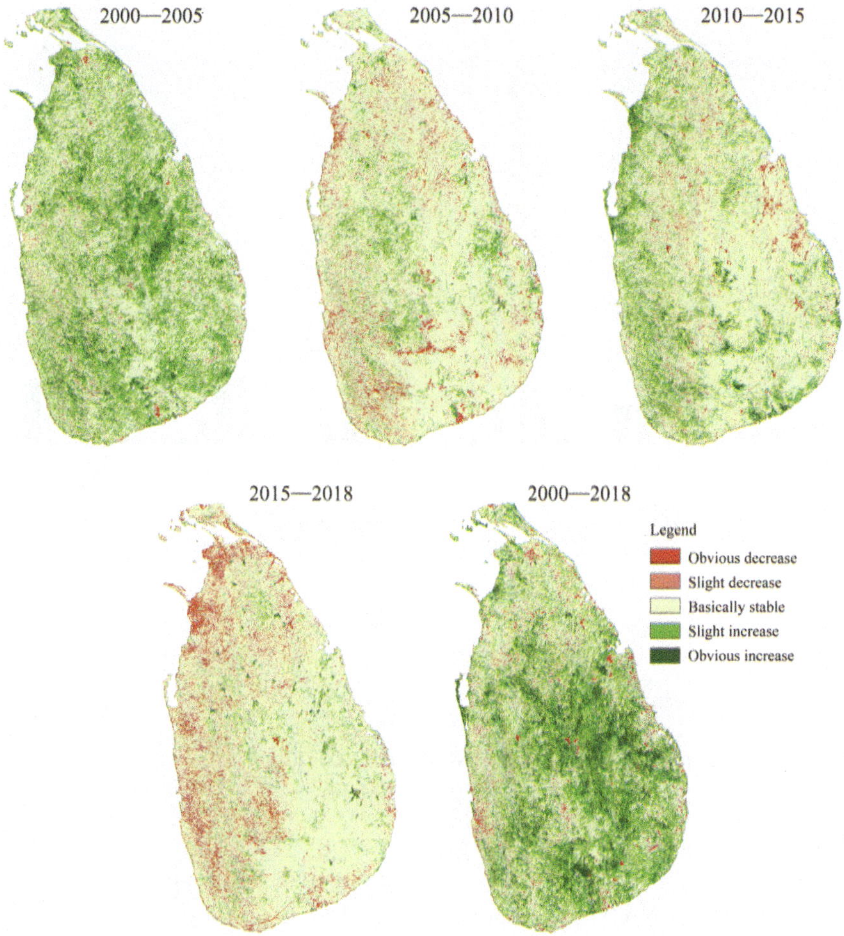

Fig. 44 Value difference of annual average NDVI 2000–2018

and low in the north. Pidurutalagala, the first peak in the country with an altitude of 2524.13 m, is located in the south of the basin. The average altitude of the whole basin is 482.2 m. The total length of the basin is about 190 km and the average width is about 60 km, with a total area of 10,448 km^2. The basin climate is a tropical monsoon climate from the middle south to the northeast, which is a transition from wet to dry. It is like summer all the year round, with the annual average temperature of 28 °C. Rainy seasons are from May to August and from November to February and the rest of the months are dry seasons. Its yearly average rainfall ranges from 1283 to 3321 mm. The upper reaches of the river are tea and rubber plantations, the middle ones are underdeveloped hilly and wet areas and the lower delta areas produce coconuts, rice, and tobacco supplemented with good irrigation equipment.

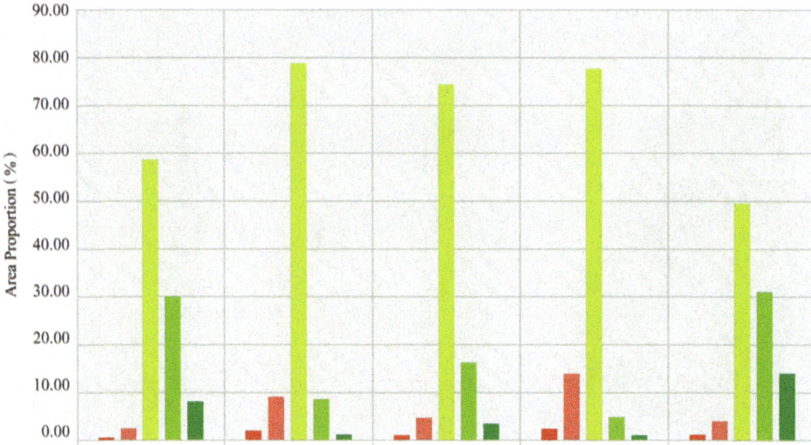

	2000—2005	2005—2010	2010—2015	2015—2018	2000—2018
obvious decrease	0.56	2.00	1.08	2.41	1.20
slight decrease	2.48	9.15	4.70	13.96	4.02
basically stable	58.75	78.91	74.46	77.71	49.59
slight increase	30.16	8.69	16.31	4.86	31.10
obvious increase	8.05	1.26	3.45	1.05	14.10

Fig. 45 Proportion of the Annual Average NDVI Area 2000–2018

Table 8 Definition of NDVI linear change trend

Slope value/year	<−0.001		−0.001~0.001	>0.001	
Significance testing ($a = 0.1$)	Yes	No		No	Yes
Variation grade	Significant decrease	Decrease	Unchanged	Increase	Significant increase

The wide-spread forest land is the largest land use in Mahaweli basin, accounting for 59.38% of the basin area slightly higher than the national proportion. Cropland is the second largest and slightly lower than the national average, mainly distributed in the high-altitude area in the basin's southwest, both sides of the north-south reach and the low-altitude area in the downstream. The water area proportion of the basin is 2.70%, significantly lower than the national average, but there are many large reservoirs and lakes for irrigation and flood regulation. The proportion of built-up land is also lower than the national average (Figs. 52, 53, 54 and 55).

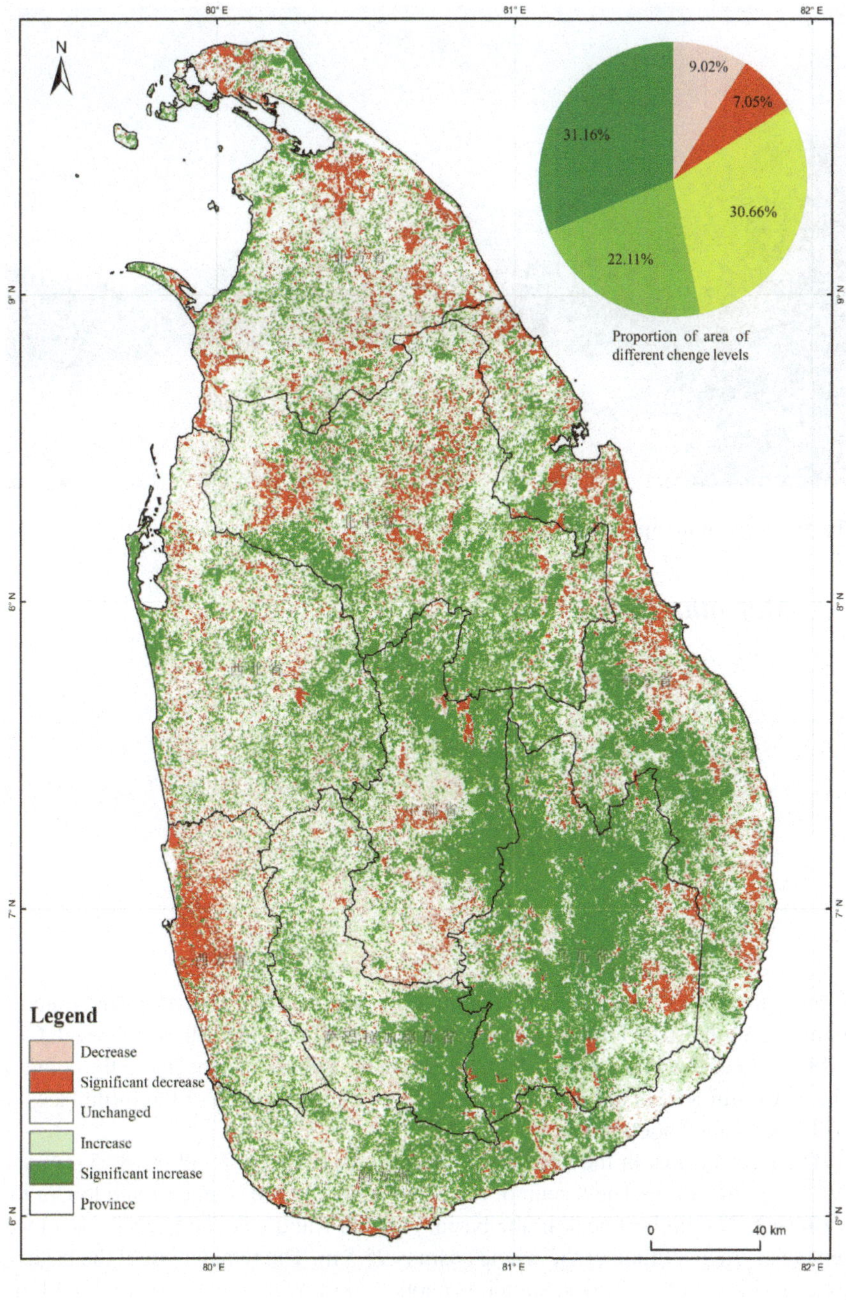

Fig. 46 Linear change trend of vegetation cover from 2000 to 2018

Fig. 47 Image of the upper reaches of Mahaweli River

3.2 *Malvathu River*

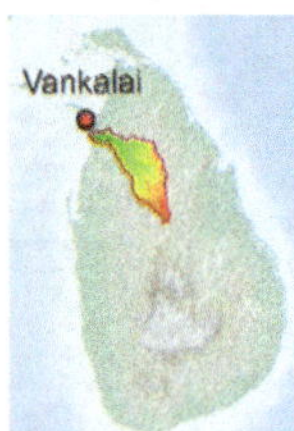

Initially known as Aruvi Aru, Malvathu is the second longest river of the country with the length of 164 km. Originating from Dambulla, it enters the sea through Gulf of Mannar near Vankalai across Anuradhapura, the largest ancient city of the country. The Malvathu River is of great historical significance and one of the birthplaces of Sri Lanka's ancient civilization.

The river locates in the northeast of the nation between 7° 58′ N–8° 52′ N and 79° 55′ E–80° 44′ E. The terrain of the basin is high in the southeast and low in the northwest. The highest peak in the basin, Ritigala Kanda, is 766 m above sea level and the average altitude of the whole basin is 95.5 m. The total length of the basin is about 125 km and the widest stretch is about 40 km, with a total area of 3284 km^2. The basin annual average rainfall is 1223 mm. The Ritigala mountain range in the upper reaches serves as an important catchment area of the Malvathu River. The Ritigala mountain range, located in the central and northern plains, 43 km southeast of Anuradhapura, is composed of four peaks, the highest one is more than 600 m

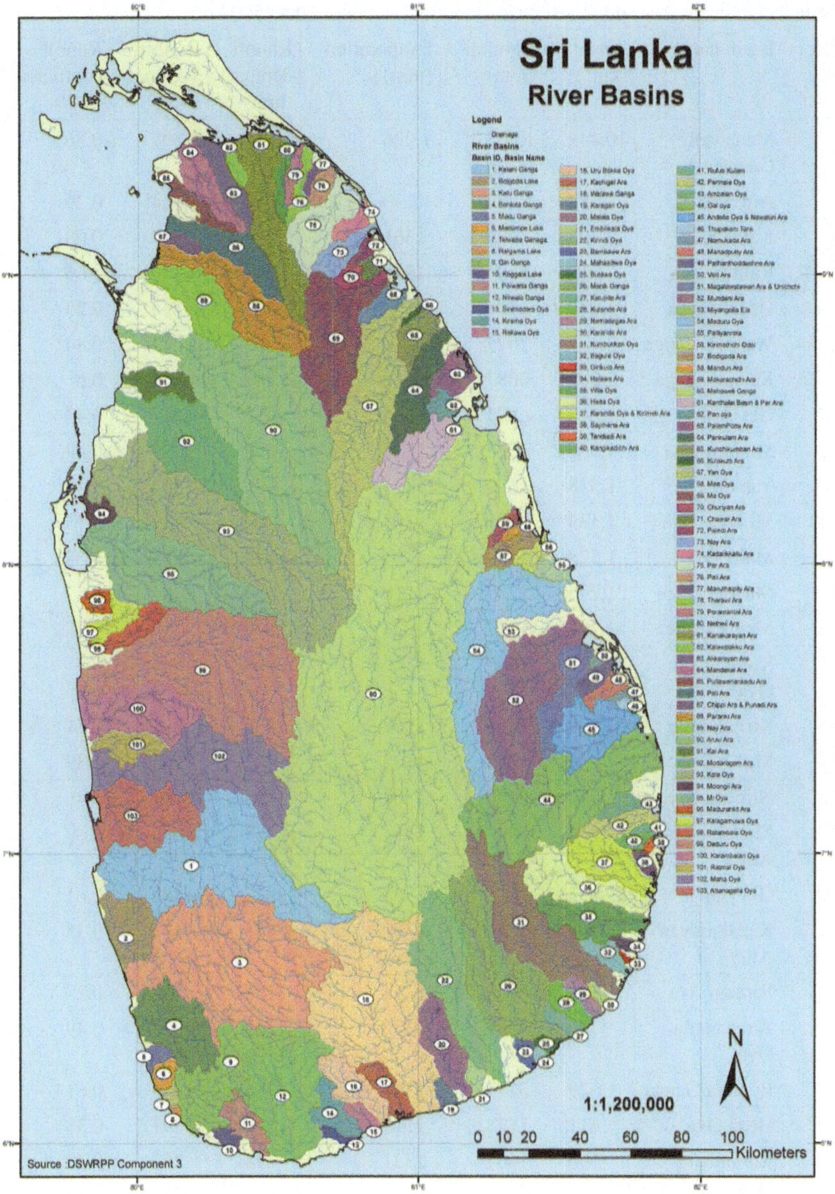

Fig. 48 Distribution of Basins. *Source* Dam Safety and Water Planning Project, Sri Lanka

Table 9 Drainage area and runoff characteristics (area more than 500 km)

No.	Basin name	Area (km^2)	Rainfall (mm)	Evaporation (mm)	Runoff depth (mm)	Runoff volume (Mm^3)	Runoff coefficient (*R*/*P*)
1	Mahaweli Ganga	10,448	1,763	1,226	641	6,697	0.36
2	Aruvi Aru	3,284	1,223	1,290	354	1,164	0.29
3	Kala Oya	2,805	987	1361	77	217	0.08
4	Kalu Ganga	2,719	3591	1315	2341	6366	0.65
5	Deduru Oya	2,647	1311	1308	300	794	0.23
6	Walawe Ganga	2,471	2077	1170	907	2242	0.44
7	Kelani Ganga	2,292	3087	1,310	1,842	4,221	0.6
8	Gal Oya	1,813	1,805	1215	782	1417	0.43
9	Maduru Oya	1,559	1573	1156	692	1079	0.44
10	Yan Oya	1,538	1629	1296	686	1054	0.42
11	Mi Oya	1,533	995	1403	104	159	0.1
12	Maha Oya	1,528	1859	1361	627	958	0.34
13	Mundeni Aru	1,295	1789	1140	850	1101	0.47
14	Menik Ganga	1,287	1417	1226	401	515	0.28
15	Kumbukkan Oya	1,233	1567	1265	475	585	0.3
16	Kirindi Oya	1,178	1485	1138	459	541	0.31
17	Ma Oya	1,036	1583	1393	622	645	0.39
18	Nilwala Ganga	971	2496	1387	1150	1117	0.46
19	Moderagam Aru	943	960	1453	84	80	0.09
20	Gin Ganga	932	3285	1333	2017	1880	0.61
21	Kanakarayan Aru	906	1299	1300	455	412	0.35
22	Parangi Aru	842	957	1353	194	164	0.2
23	Attanagalla Oya	736	2595	1385	1260	927	0.49
24	Bentara Ganga	629	3271	1348	1988	1250	0.61
25	Heda Oya	611	1678	1197	640	391	0.38
26	Karambala Oya	596	1461	1331	473	282	0.32
27	Nay Aru	567	1171	1388	305	173	0.26
28	Andella Oya	528	2212	1151	1186	626	0.54

Source W.G.M Bastiaanssen, L Chandrapala. Water balance variability across Sri Lanka for assessing agricultural and environmental water use. Agricultural Water Management, 2003, 58(2): 171–192

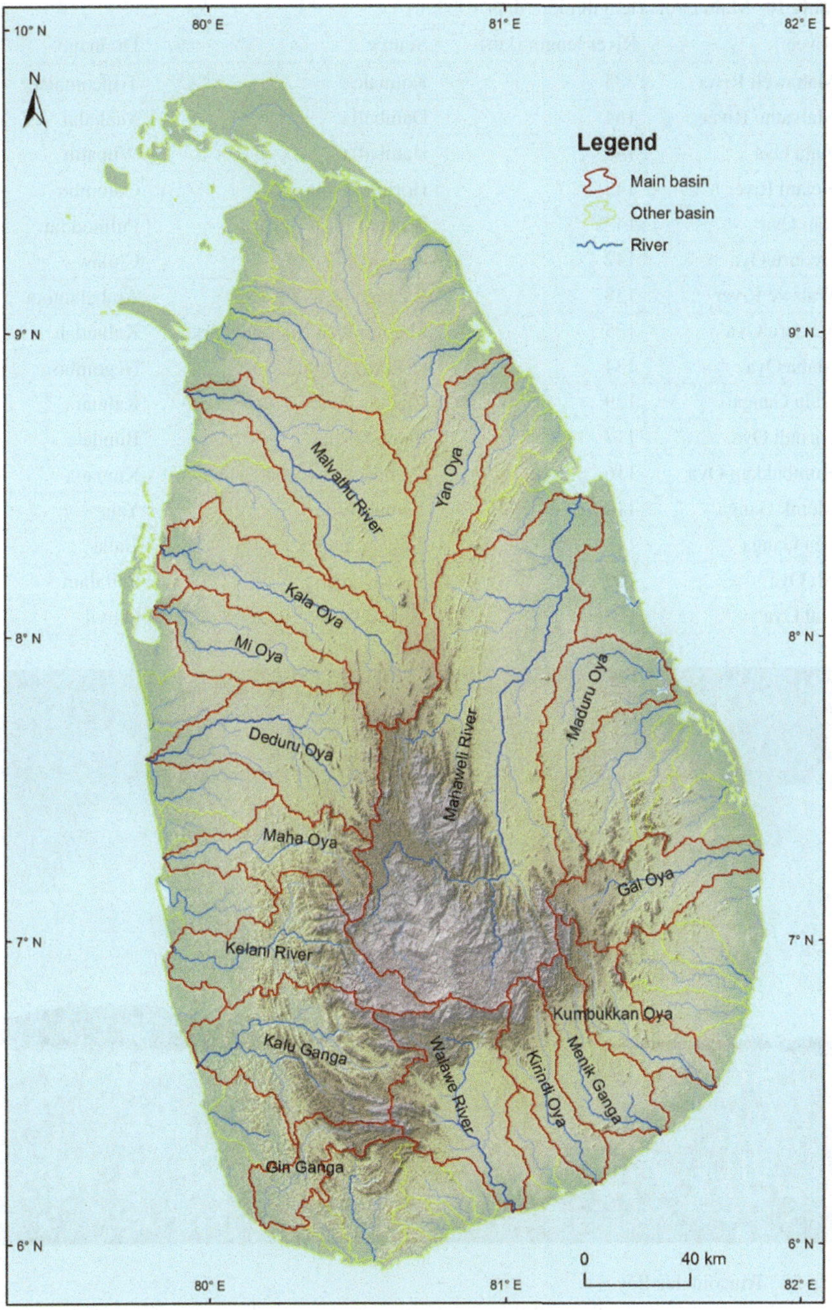

Fig. 49 Distribution image of main rivers (length more than 100 km)

Table 10 Main rivers (length more than 100 km)

River	River length (km)	Source	Drainage
Mahaweli River	335	Kotmale	Trincomalee
Malvathu River	164	Dambulla	Vankalai
Kala Oya	148	Dambulla	Wilpattu
Kelani River	145	Horton Plains	Colombo
Yan Oya	142	Ritigala	Pulmoddai
Deduru Oya	142	Kurunegala	Chilaw
Walawe River	138	Balangoda	Ambalantota
Maduru Oya	135	Maduru Oya National Park	Kalkudah
Maha Oya	134	Hakurugammana	Negombo
Kalu Ganga	129	Adam's Peak	Kalutara
Kirindi Oya	117	Bandarawela	Bundala
Kumbukkan Oya	116	Dombagahawela	Kumana
Menik Ganga	114	Namunukula	Yala
Gin Ganga	113	Sinharaja	Galle
Mi Oya	109	Saliyagama	Puttalam
Gal Oya	108	Yalkumbura	Oluvil

Fig. 50 Trincomalee Bay

Fig. 51 Remote sensing image of Trincomalee Bay in January 2017

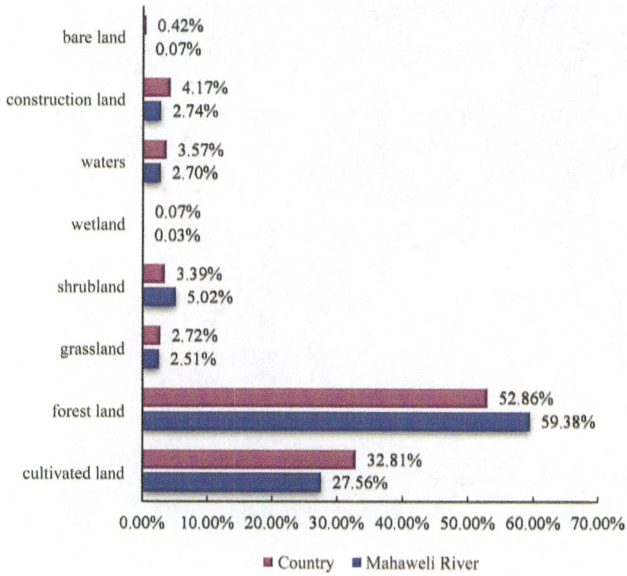

Fig. 52 Comparison of land use between Mahaweli River and the country

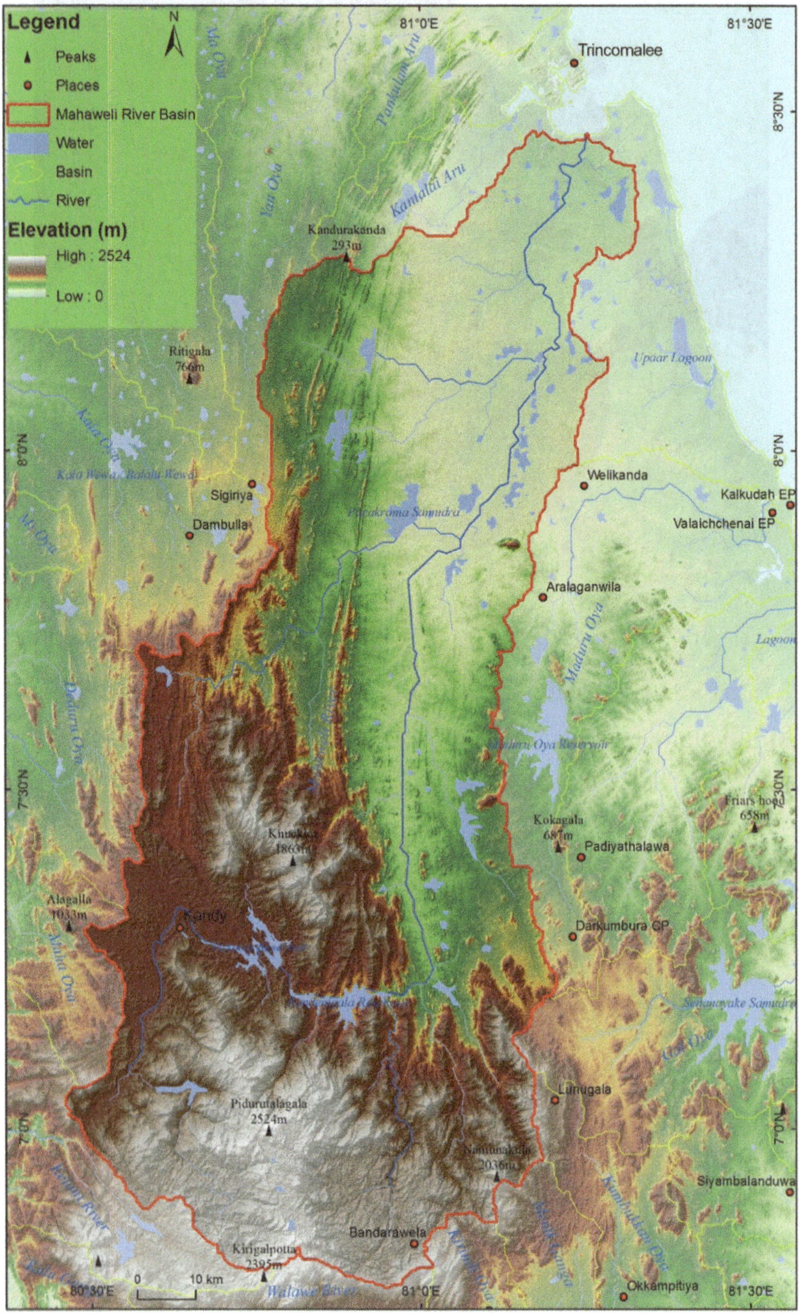

Fig. 53 Image of Mahaweli River

Fig. 54 True color remote sensing image of Mahaweli River

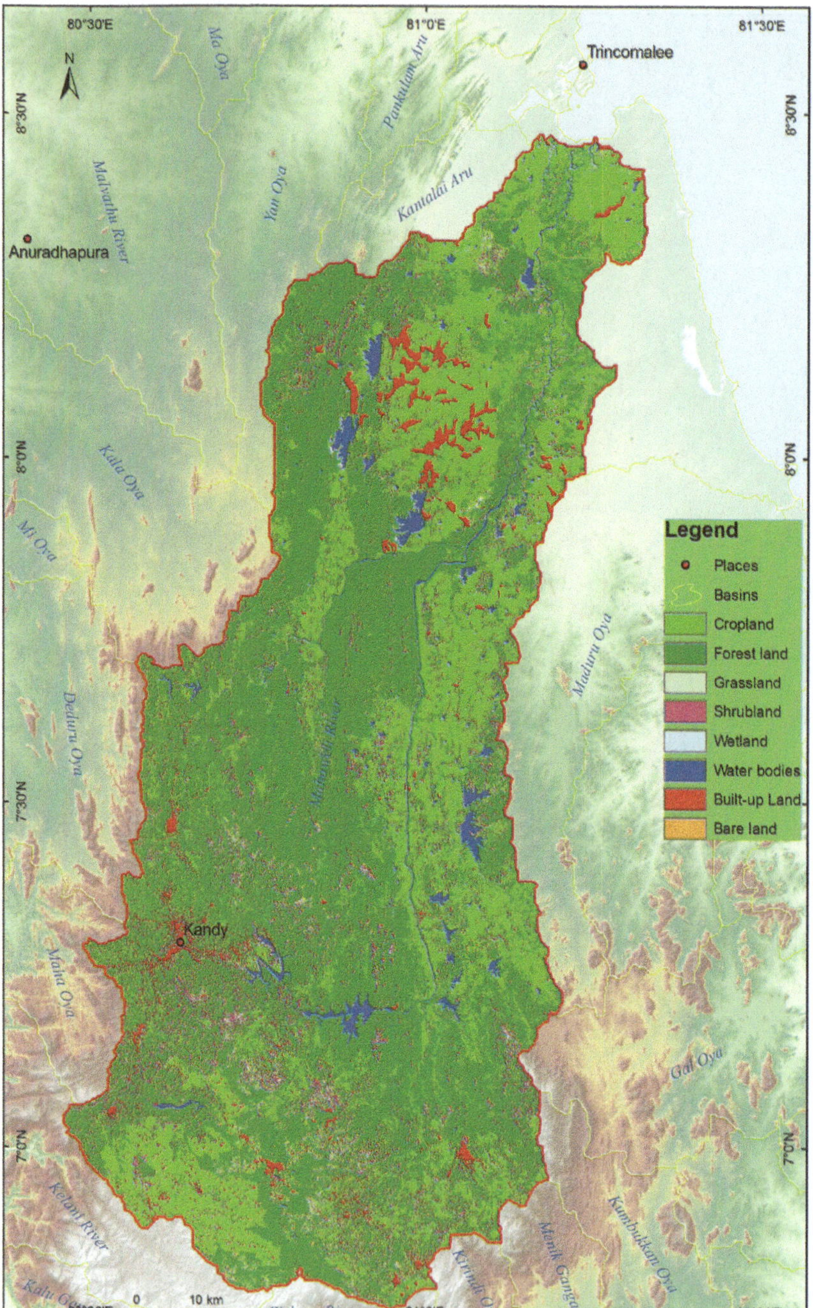

Fig. 55 Land classification of Mahaweli River by remote sensing

higher than the surrounding plain, making it the highest peak in northern Sri Lanka. Steep with luxuriant trees, the mountain range is the most abundant rainfall area and an important water conservation in the northern arid region. It is also the catchment area of the Malvathu River and Yan Oya, consequently building the Ritigala Strict Nature Reserve.

Mostly in the basin's middle and upper reaches, the cropland area accounts for 52.90% of the total basin area, significantly higher than that of the country's average. The forest land is mainly distributed in southward the Maduroad and Mavillu forest reserves in the lower reaches as well as hilly areas in the river origin and the farmland of the middle and upper reaches. The proportion of forest land area is significantly lower than that of the national forest land area. Moreover, the water area ratio is slightly lower than the national average, but can meet agricultural irrigation and domestic water demand in the agricultural planting area, the reservoir and ponds. Further, the proportion of built-up land area is significantly higher than the national average (Figs. 56, 57, 58 and 59).

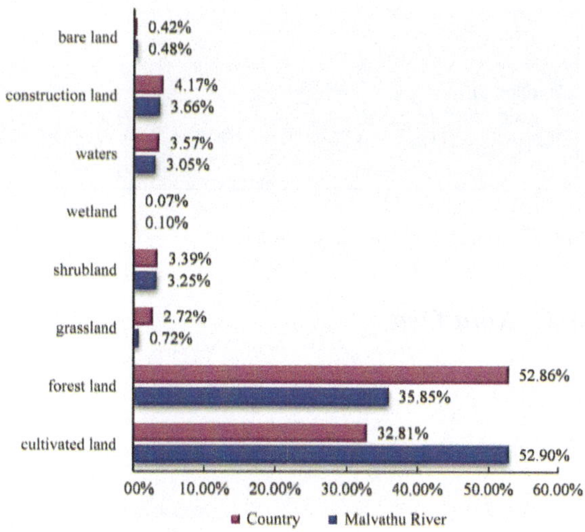

Fig. 56 Comparison of land use between Malvathu River and the country

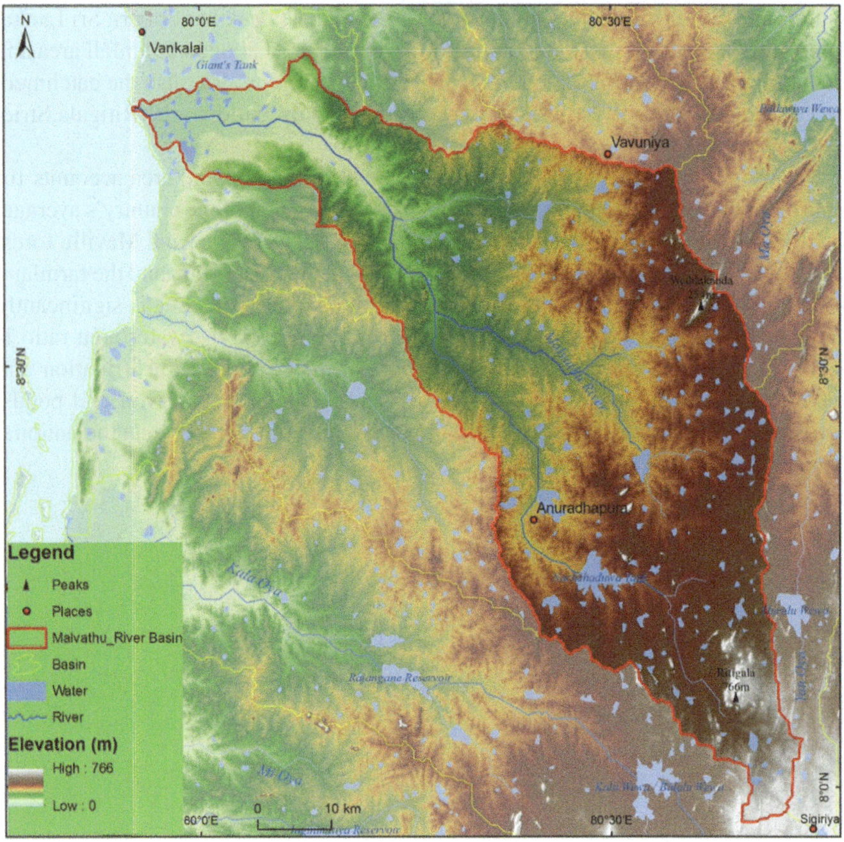

Fig. 57 Topography of Malvathu River

3.3 *Kala Oya*

It is the third longest river across the country with a length of 148 km. Originating in the hilly areas of Dambulla, and flows from southeast to northwest and then enters the Dutch Bay of Indian Ocean near Wilpattu.

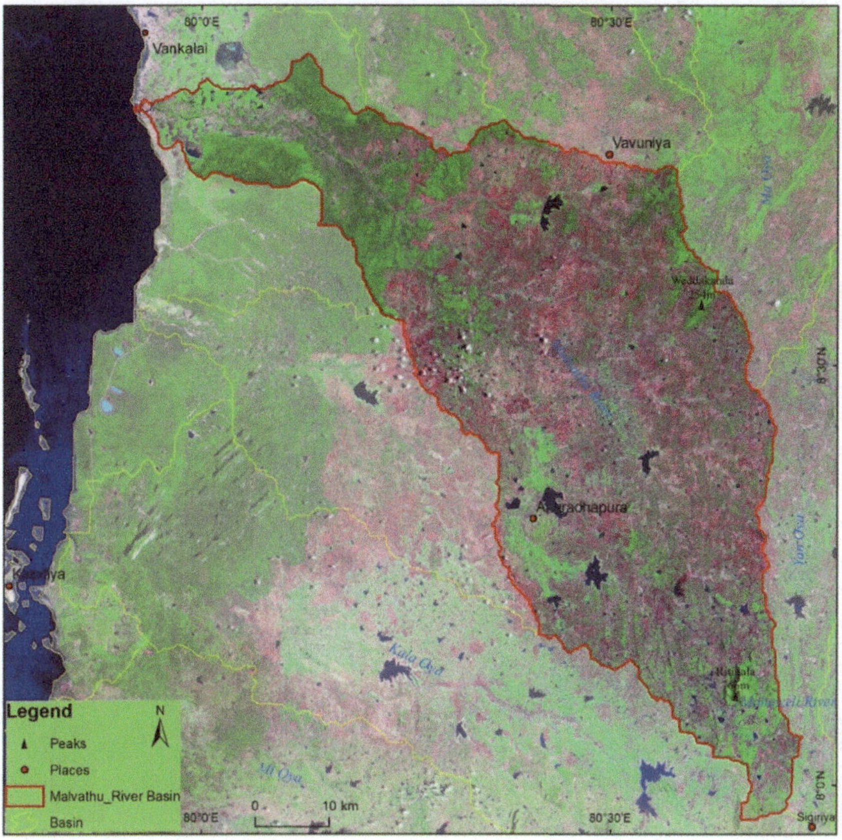

Fig. 58 True color remote sensing image of Malvathu River

Kala Oya is located in the north between 7° 42′ N–8° 27′ N and 79° 49′ E–80° 47′ E. The terrain of the basin is high in the southeast and low in the northwest. It peaks at 868 m above sea level and an average basin elevation of 120.1 m. The total length is about 115 km and the average width is about 22 km, with a total area of 2870 km^2. The basin receives about 3.169 billion cubic meters of rainfall every year, of which about 5% flows into the sea.

Mostly in the middle and upper reaches, the cropland area accounts for nearly 40% of the total basin area, higher than national average. The forest land is mainly distributed in the south of the Wilpattu National Park and north of Tabbowa Wildlife Sanctuary in the lower reaches as well as some hilly areas in the river origin, upper and middle reaches. The proportion of forest land area is significantly lower than the national average. Also, the water area ratio is slightly lower than the national average, but can still meet agricultural irrigation and domestic water demand in the

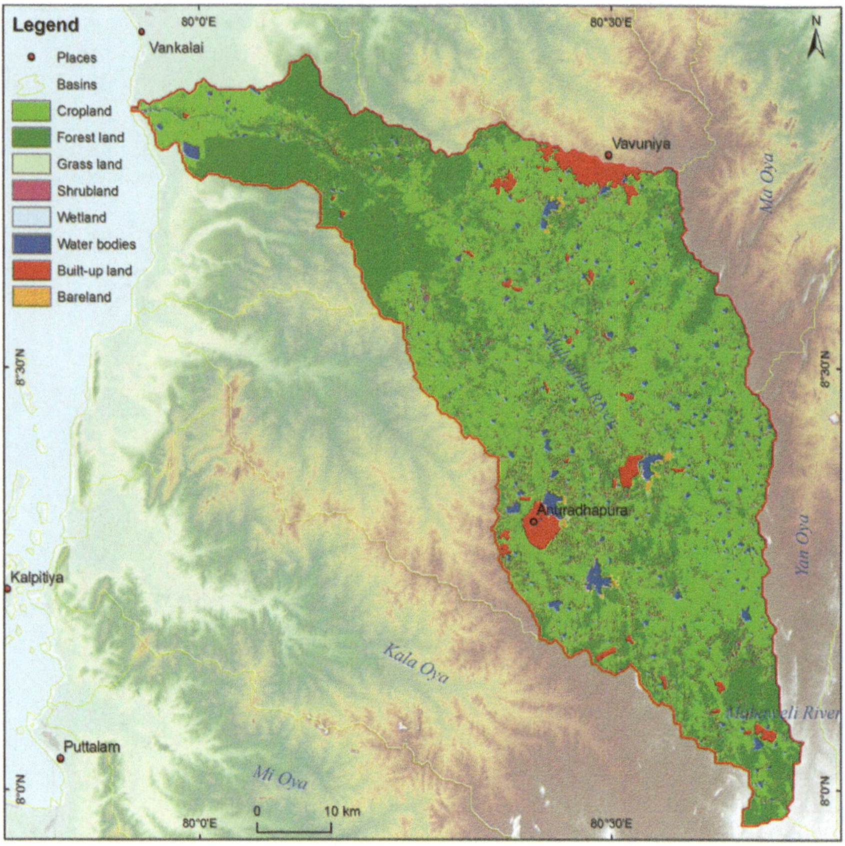

Fig. 59 Land Classification of Malvathu River by remote sensing

agricultural planting area, the reservoir and ponds. Furthermore, the proportion of built-up land area is significantly higher than the national average (Figs. 60, 61, 62 and 63).

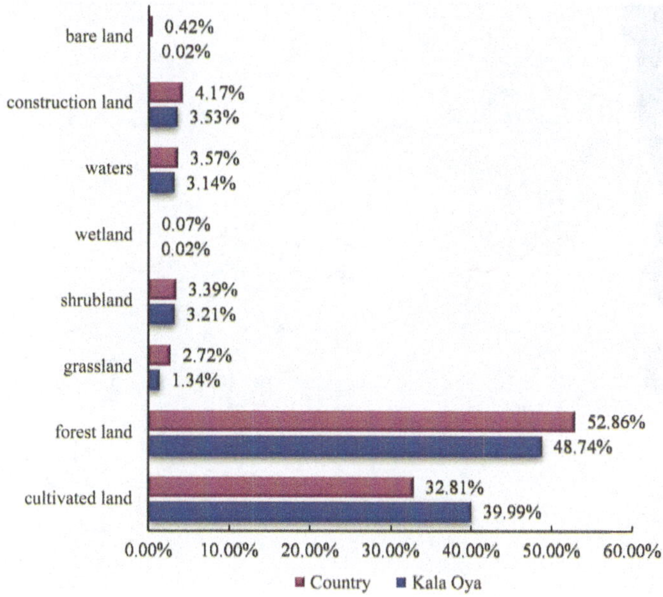

Fig. 60 Comparison of land use between Kala Oya and the country

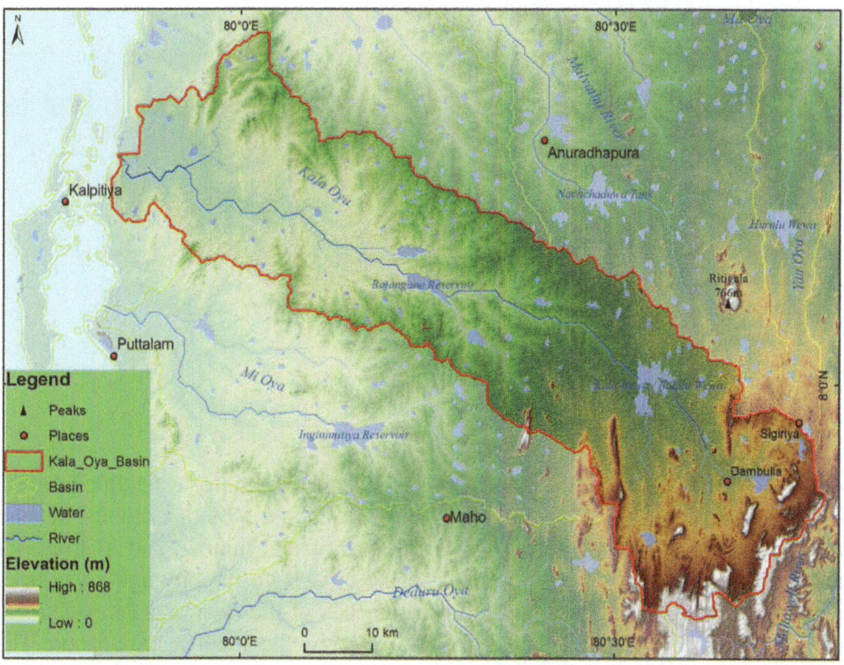

Fig. 61 Topography of Kala Oya

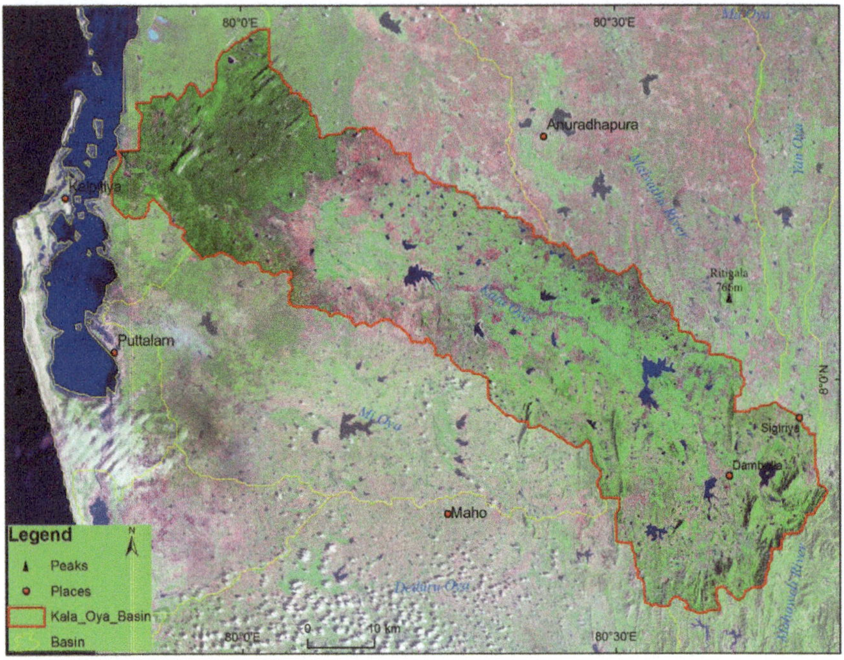

Fig. 62 True color remote sensing image of Kala Oya

3.4 *Kelani River*

With a length of 145 km, it is the fourth longest river in the country, extending from the Sri Pada Mountain Range to Colombo. It provides about 80% of Colombo's water used for transportation, fishing, sewage treatment, sand mining, and hydropower generation but also poses a flood risks to the city. Both the city and suburbs of Colombo located in the plain at the lower reaches of the river and the catchment area at the upper reaches of the river are affected by the southwest monsoon. The water level in rainy season corresponds to the flood risks in the capital. Another major problem is the saltwater intrusion in the dry season, which may become more serious due to the deepening of the river caused by sand mining.

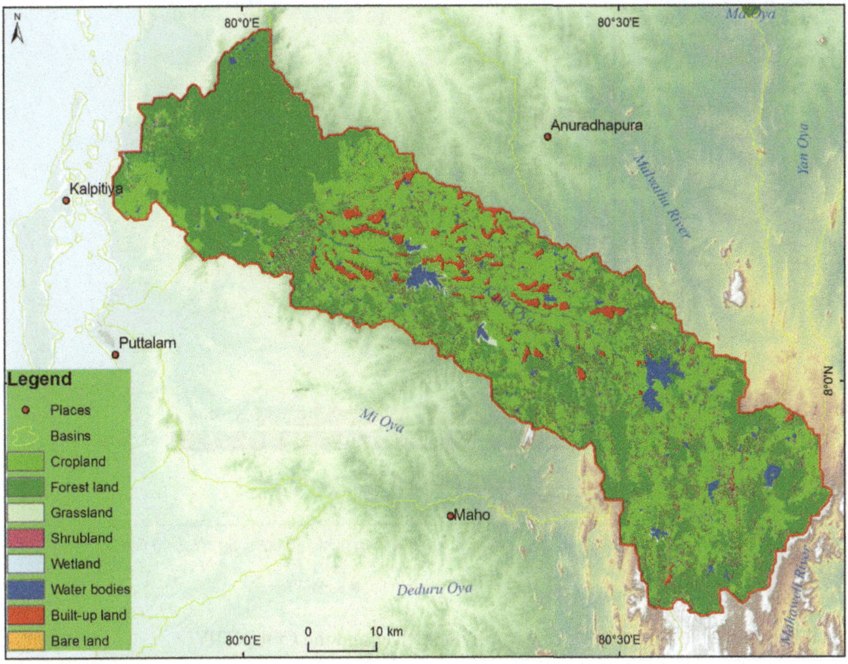

Fig. 63 Land classification of Kala Oya by remote sensing

Kelani River is located in the southwest between 6° 45′ N–7° 14′ N and 79° 52′ E–80° 46′ E. The terrain of the basin is high in the East and low in the West and the average elevation of the whole basin is 372.6 m. The highest peak in the basin, Kirigalpotta, is 2,395 m above sea level, serving as the second highest peak in Sri Lanka. The total length of the basin is about 102 km and the widest stretch is about 45 km, with a total area of 2,294 km^2. The annual average rainfall there is 3,087 mm, offering adequate water supply.

The area proportion of built-up land mostly in the downstream is significantly higher than the national average, 2.81 times of the national proportion. Also, cropland area mainly in the upstream and downstream accounts for nearly 16.85% of the total basin area, half as much as the national cropland proportion. The forest land area proportion is higher than the national average, mainly concentrated in the middle and upper reaches including the northwest of the Peak Wilderness Protected Area. The water area rate (1.23%) is significantly lower than the national average. There are few large reservoirs except for two in the upper reaches, namely Maussakelle reservoir and Castlereigh reservoir (Figs. 64, 65, 66 and 67).

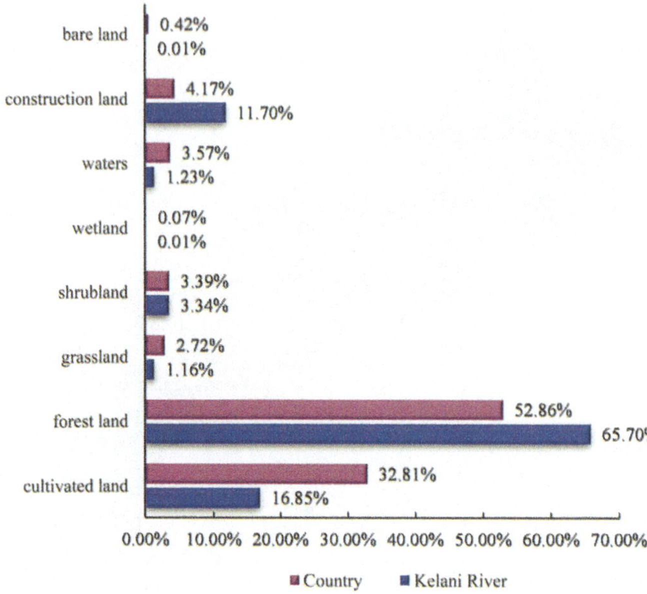

Fig. 64 Comparison of land use between Kelani River and the COUNTRY

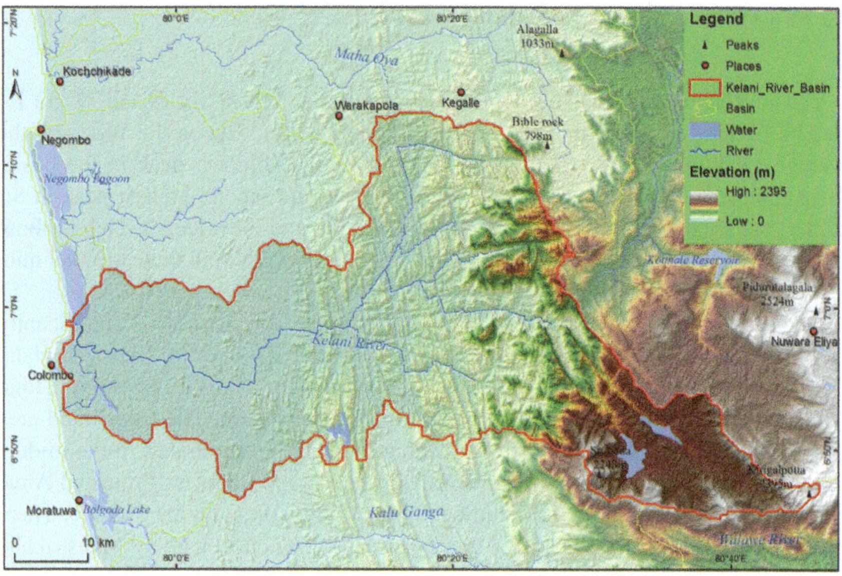

Fig. 65 Topography of Kelani River

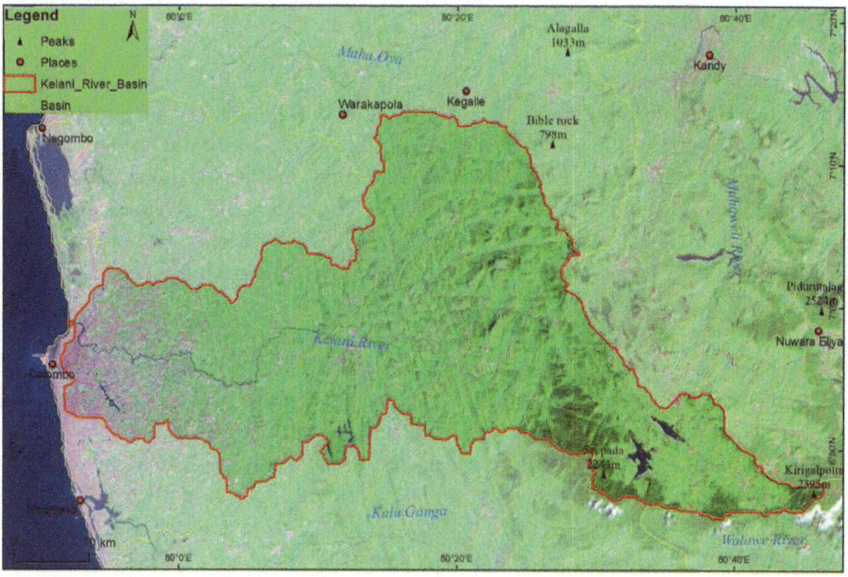

Fig. 66 True color remote sensing image of Kelani River

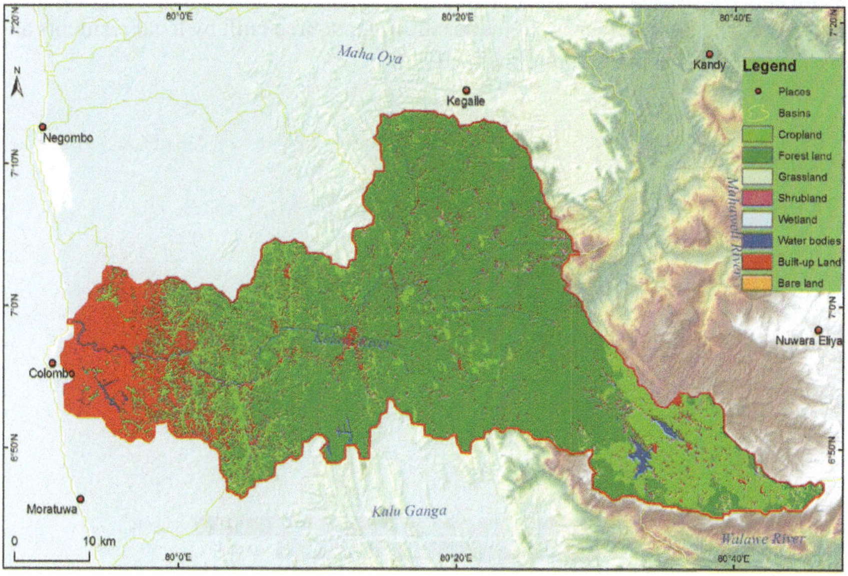

Fig. 67 Land classification of Kelani River by remote sensing

3.5 *Yan Oya*

Originating from the mountainous and hilly areas of Dambulla and Sigiriya, Yan Oya flows from the southwest to the northeast. It is incorporated into two tributaries, Adappan Oya and Sellige Oya in the middle reaches, and it is also known as Sigiri Oya (Habarana Oya) in the upper reaches. It finally enters the Indian Ocean in the south of Pulmoddai with a tributary, Thawalam Halmilla Oya in the Pangurugaswawa village of the lower reaches. Its main stream is about 142 km long, with a constant cross-section and a gradient of 1.5–2%.

The river is located in the northeast of Sri Lanka between 7° 55′ N–8° 56′ N and 80° 41′ E–81° 02′ E. The basin is characterized with high terrain in the southwest and low in the northeast. It peaks at 410 m above sea level and a basin average elevation of 92.0 m. The total length is about 120 km and the average width is about 15–20 km, with a total area of 1538 km^2. Located in the tropical marine climate zone, it has rainy season from October to February and thereafter dry seasons with yearly average annual rainfall of 1629 mm. Yan Oya has witnessed sound social and economic development in the densely populated areas of its upper reaches, especially Hurulu Wewa. However, the lower reaches, which are lowland plain, are mostly covered with shrubs and have not been reclaimed. There are two large reservoirs (Hurulu Wewa and Wahalkada), plenty of ponds, and ancient irrigation canals that have been renovated. The pond serves as an important water source built by local residents and have villages distributed around them (Fig. 68).

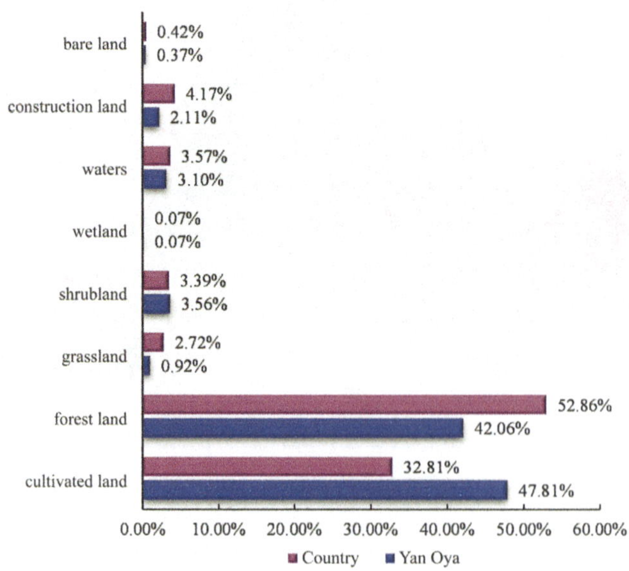

Fig. 68 Comparison of land use between Yan Oya and the country

The proportion of cropland along Yan Oya accounts for nearly 47.81% of the total basin, 1.46 times of the national average. As for forest land mainly distributed in the lower reaches and the east of the middle reaches, its proportion is lower than the national average. The water area coverage is slightly lower than the national average, but they are all fresh water used for agricultural irrigation and domestic water. Despite being located in the arid land, the densely distributed ponds and large reservoirs compliment agricultural use demands. Further the proportion of built-up land area is half of the national average (Fig. 69, 70 and 71).

3.6 Deduru Oya

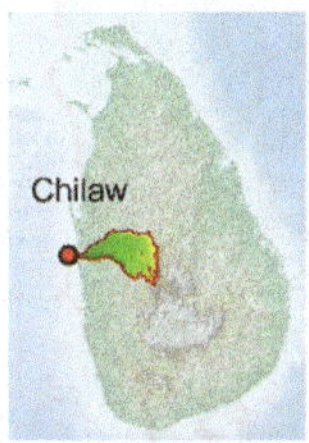

About 142 km long, it originates from hills around Matale and enters the Indian Ocean around Chilaw.

It is located in the northwest between 7° 20′ N–7° 51′ N and 79° 47′ E–80° 35′ E. The basin is characterized by high terrain in the east and low in the west peaking at 1255 m above sea level with an average basin altitude of 127.5 m. The total length is about 87 km and the widest stretch is about 55 km, with a total area of 2647 km^2. The total population along the river is about 1 million.

The annual average rainfall is 1628 mm and the evaporative capacity is 1250 mm. The proportion of water for irrigation and agriculture is fairly high compared with that for domestic, industry and commerce. It has four main reservoirs to support the irrigation agricultural system as well as about 3000 small and medium-sized reservoirs and ponds for irrigation and domestic water.

The cropland is widely distributed in the basin and the area proportion is larger than the national average. The forest land, mainly in the east of the upstream and the north-west edge of the Central Highlands, is slightly less than the national average area. The water area coverage is about half of the national average and the built-up land proportion is significantly lower than the national average (Figs. 72, 73, 74 and 75).

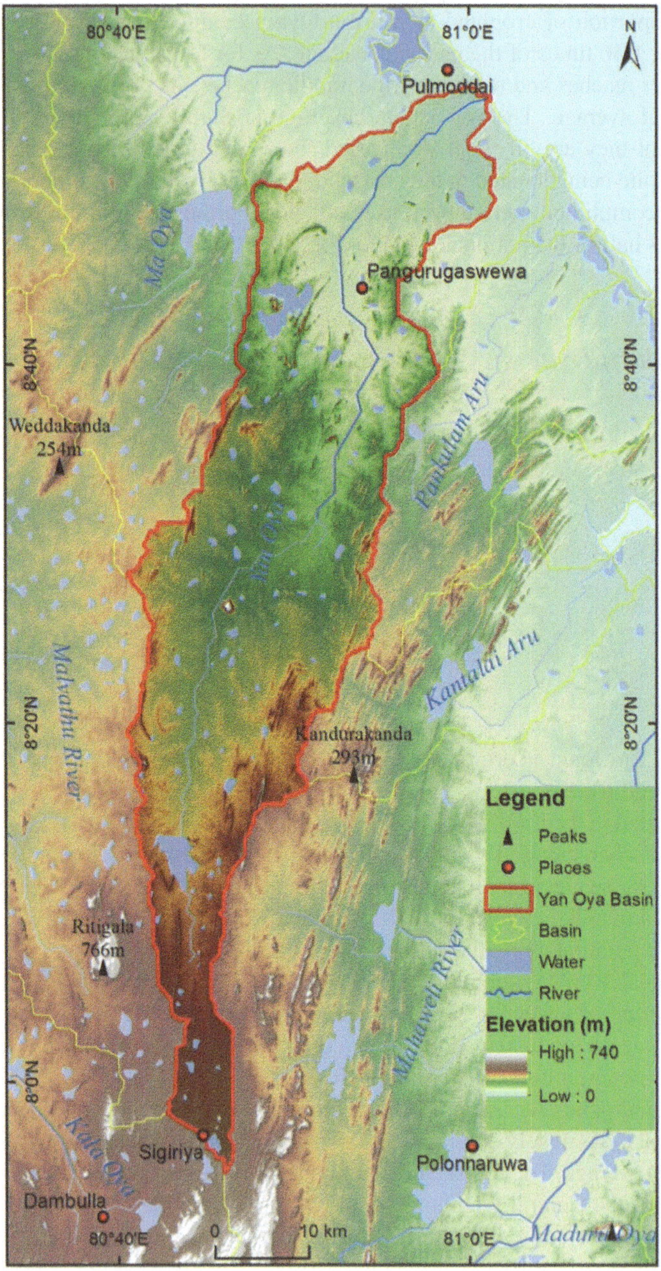

Fig. 69 Topographic of Yan Oya

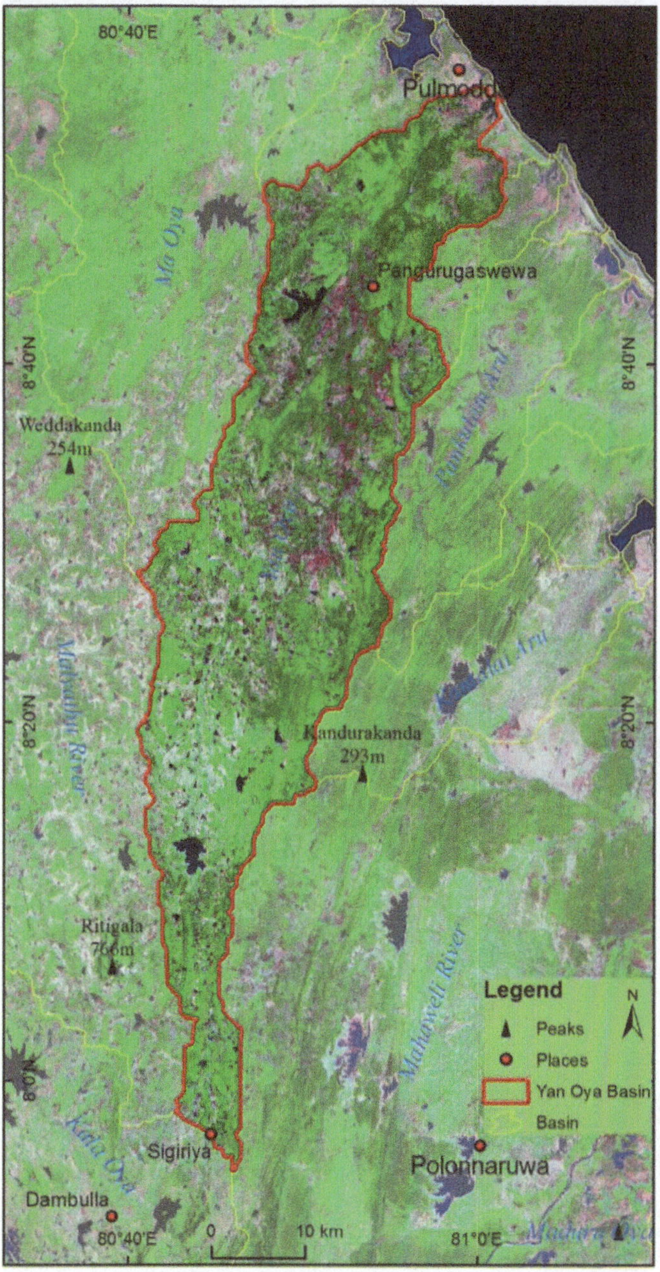

Fig. 70 True Color Remote Sensing Image of Yan Oya

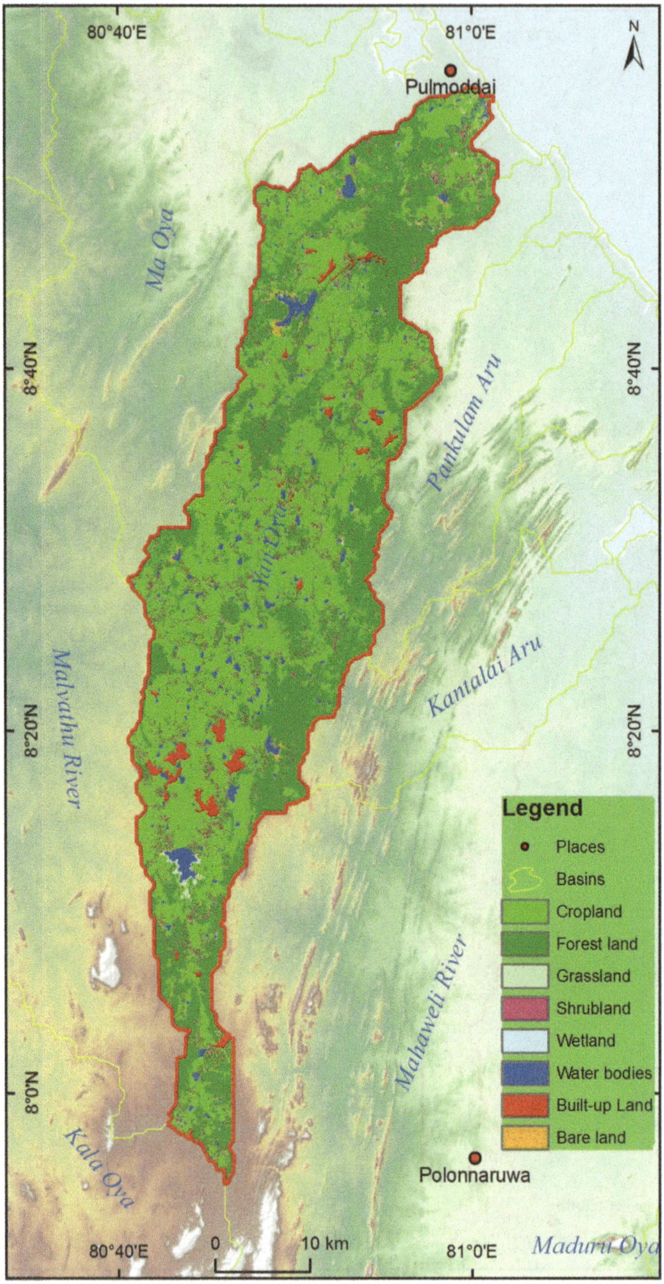

Fig. 71 Land classification of Yan Oya by remote sensing

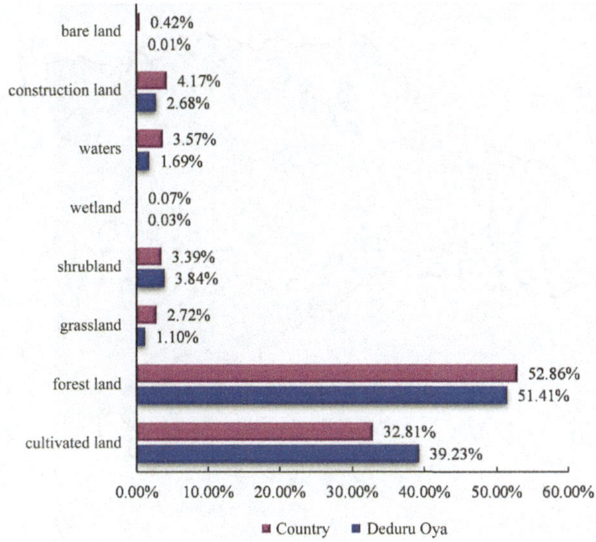

Fig. 72 Comparison of land use between Deduru Oya and the country

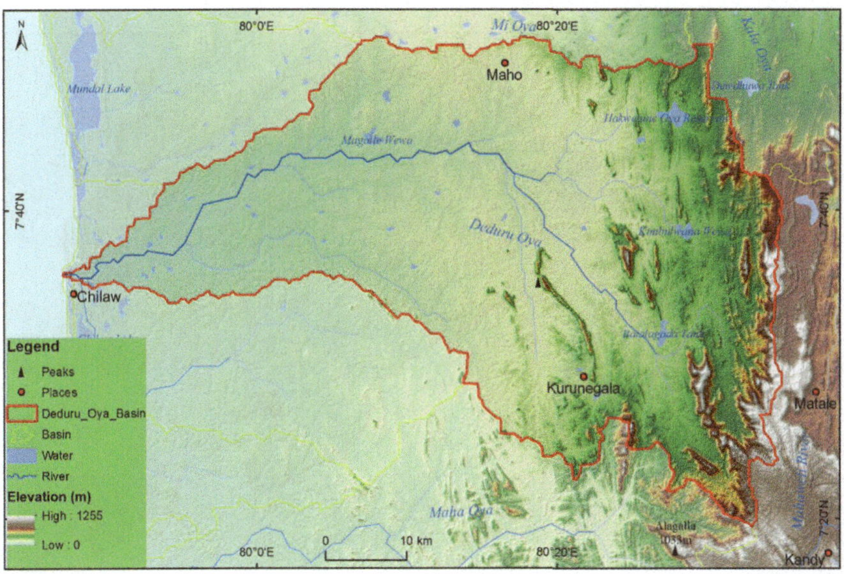

Fig. 73 Topography of Deduru Oya

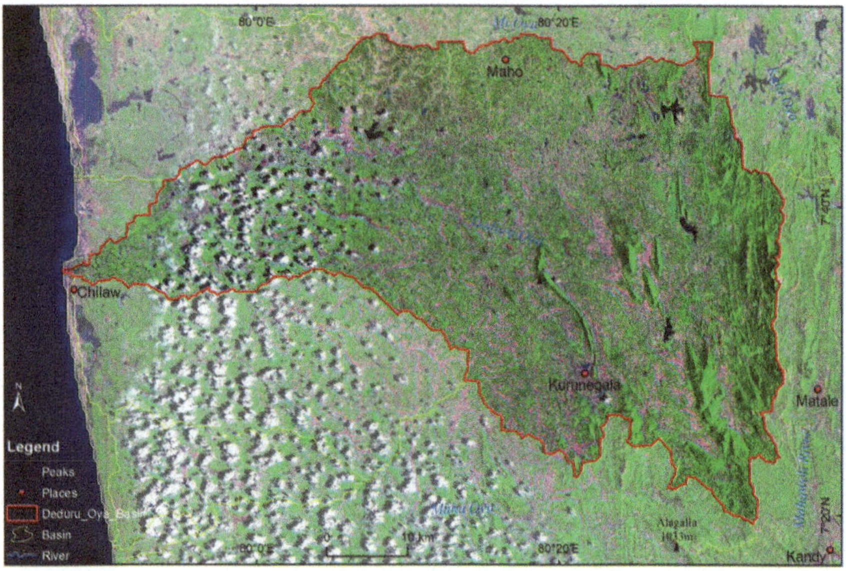

Fig. 74 True color remote sensing image of Deduru Oya

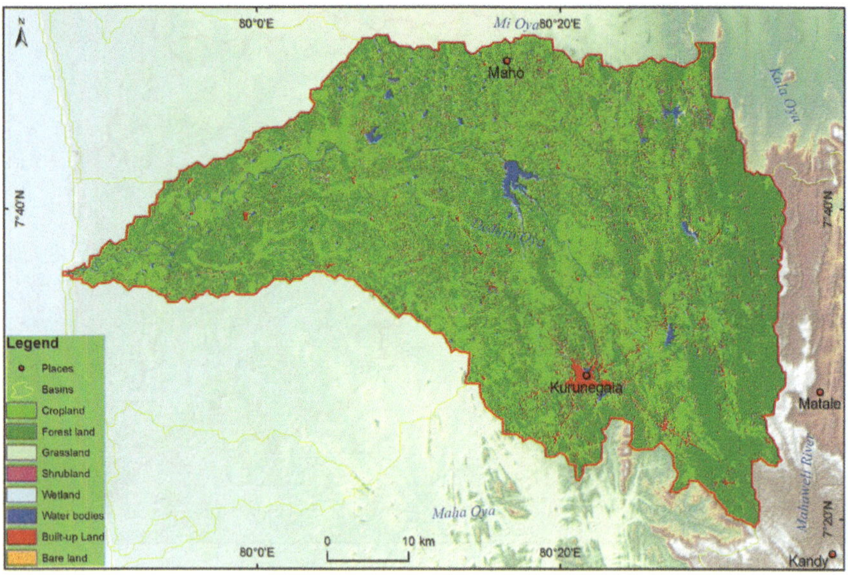

Fig. 75 Land classification of Deduru Oya by remote sensing

3.7 *Walawe River*

About 138 km long, Walawe River originates from Adam's Peak in the Peak Wilderness Protected Area and flows from north to south into the Indian Ocean near Ambalantota.

It is located in the southwest between 6° 06′ N–6° 49′ N and 80° 33′ E–81° 05′ E. The basin is characterized by high terrain in the north and low in the south peaking at 2234 m above sea level with basin average altitude of 348 m. The total length is about 83 km and the widest stretch is about 50 km, with a total area of 2471 km^2. There are mountains and hills in the north and west, and plains in areas below 200 m above sea level in the middle and south, mainly cropland.

The cropland mostly in the south and northeast of Udawalawa National Park accounts for nearly 28.69% of the total basin area, which is slightly lower than the national average. The proportion of forest land area is slightly higher than the corresponding national average, mainly distributed in the west, north of the middle reaches and the whole upstream. The water area coverage is slightly lower than the national average, and the proportion of built-up land area is three-quarters of the national average proportion, mainly distributed in the downstream (Figs. 76, 77, 78 and 79).

3.8 *Maduru Oya*

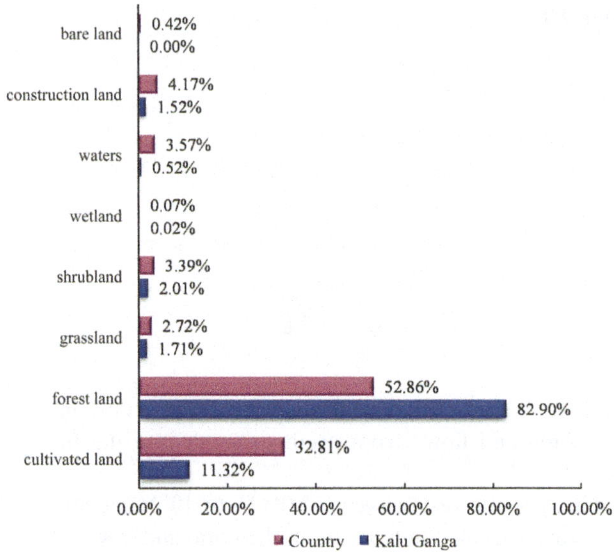

Fig. 76 Comparison of land use between Walawe River and the country

About 135 km long, Maduru Oya originates from the Maduru Oya National Park and flows from south to northeast into the Indian Ocean near Kalkudah. The Maduru Oya National Park built in Novermber 1983 is a crucial water source for Maduru Oya.

The river locates in the eastern central region between 7° 12′ N–8° 00′ N and 81° 07′ E–81° 33′ E. The basin is characterized by high terrain in the south and low in the north and east peaking peaks at 858 m above sea level with basin average altitude of 82.7 m. The total length is about 90 km and the widest width is about 30 km, with a total area of 1559 km^2. The south of the basin is mountainous and hilly while north and eastern areas are plains below 200 m above sea level dominated by cropland.

Mostly distributed in the middle and low reaches, the cropland accounts for nearly 40.36% of the total basin area, which is higher than national average. The forest land is mainly distributed in the downstream Thoppigala Forest Reserve and upstream Maduru Oya National Park as well as some hilly areas in the river profile with its proportion little smaller than the national average. Moreover, the water area coverage is 1.38 times the national average contributed significantly by Maduru Oya Reservoir in the basin's upper reaches, while the proportion of built-up land area is significantly lower than the national average (Figs. 80, 81, 82 and 83).

Fig. 77 Topography of Walawe River

3.9 *Maha Oya*

About 134 km long, it originates in the northwest edge of the Central Highlands and flows from southeast to northwest and later flows westward into the Indian Ocean near Negombo.

It is located in the middle east of Sri Lanka between 7° 04′ N–7° 29′ N and 79° 51′ E–80° 33′ E. The basin's terrain is high in the east and low in the west peaking at 1,498 m above sea level with an average altitude of 184.4 m. The total length is about 78 km and the average width is about 20–30 km, with a total area of

Fig. 78 True color remote sensing image of Walawe River

1528 km^2. The middle east of the basin is mountainous and hilly and gently sloping in the west.

The cropland mainly distributed in the middle and lower reaches of Maha Oya accounts for nearly 20.31% of the total basin area, which is about two-thirds of the national average cropland area. The proportion of forest land is higher than the national average, about 1.33 times of the national average proportion. The middle and upper reaches are dominated by forest land while the lower reaches are evenly distributed. The basin has relatively abundant rainfall, and without large and medium-sized reservoirs. The water area proportion is very low, less than one fifth of the

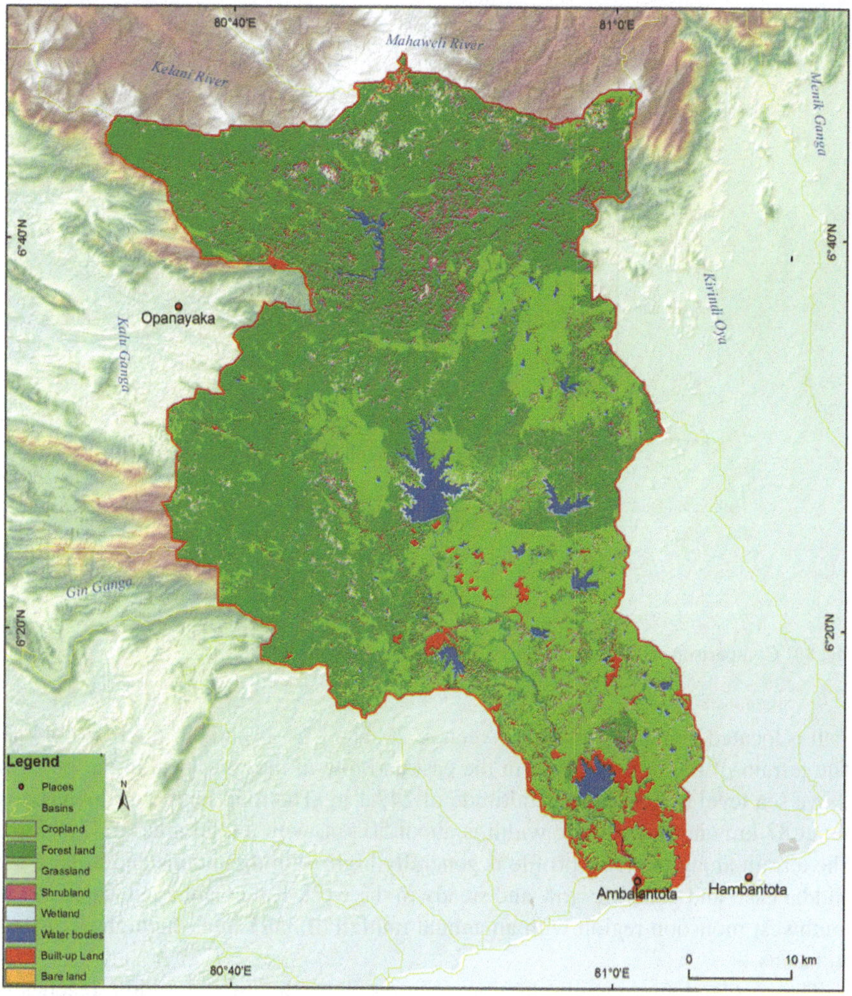

Fig. 79 Land classification of Walawe River by remote sensing

national average and the proportion of built-up land area is also lower than national average (Figs. 84, 85, 86 and 87).

3.10 Kalu Ganha

About 129 km long, it originates from the Adam's Peak in the Central Highlands fed by the mountain forest and Sinharaja Forest Reserve and flows from east to west entering Indian Ocean around Kalutara.

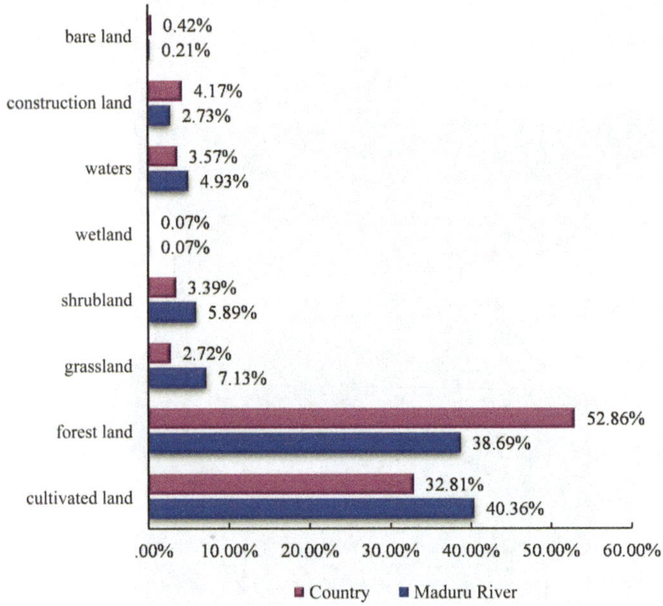

Fig. 80 Comparison of land use between Maduru Oya and the country

It is located in the southwest between 6° 20′ N–6° 54′ N and 79° 55′ E–80° 44′ E. The terrain of the basin is high in the east and low in the west peaking at 2073 m above sea level and an average altitude of 247.1 m. The total length of the basin is about 87 km and the average width is about 50 km, with a total area of 2719 km^2. The terrain along the river profile is generally high, with mountains and hills in the middle east, and generally low and steady in the west. Kalu Ganha is located in the southwest monsoon region with an annual rainfall of 3591 mm, the highest among all basins.

The cropland area mainly distributed in the downstream and middle reaches is as small as 11.32%, which is one third of the national average cropland area. The forest land area occupies 82.90% of the whole basin, widely distributed in the whole basin. The basin has abundant rainfall, few reservoirs and ponds, and the water area coverage is about 0.52%, far lower than the national average. The built-up land area is also lower than the national average, mainly distributed downstream (Figs. 88, 89, 90 and 91).

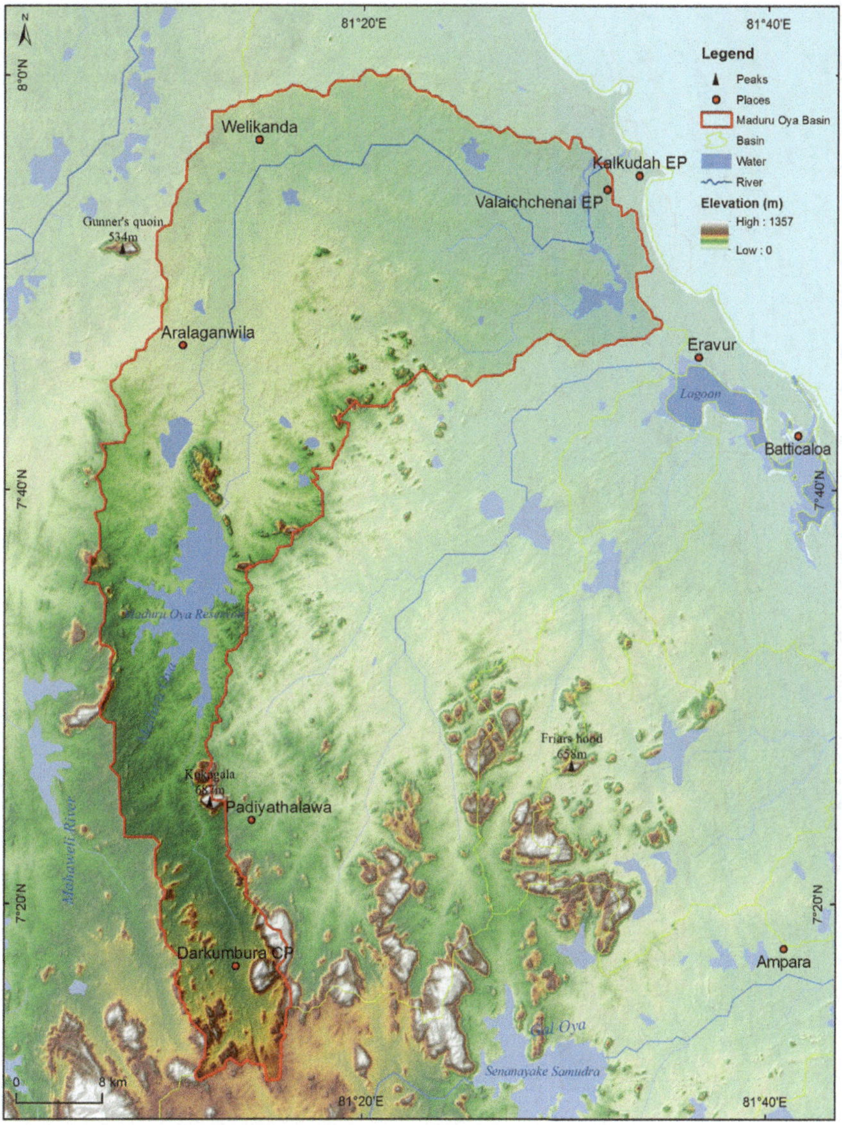

Fig. 81 Topography of Maduru Oya

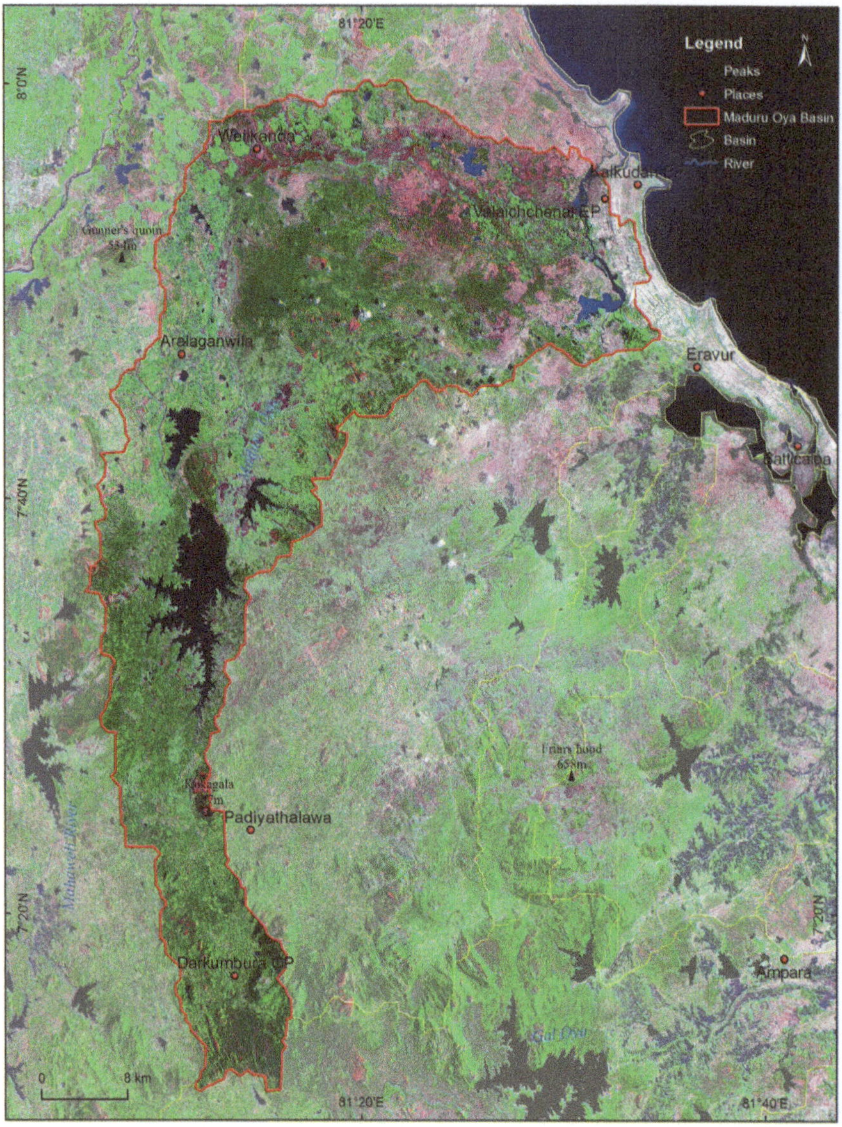

Fig. 82 True color remote sensing image of Maduru Oya

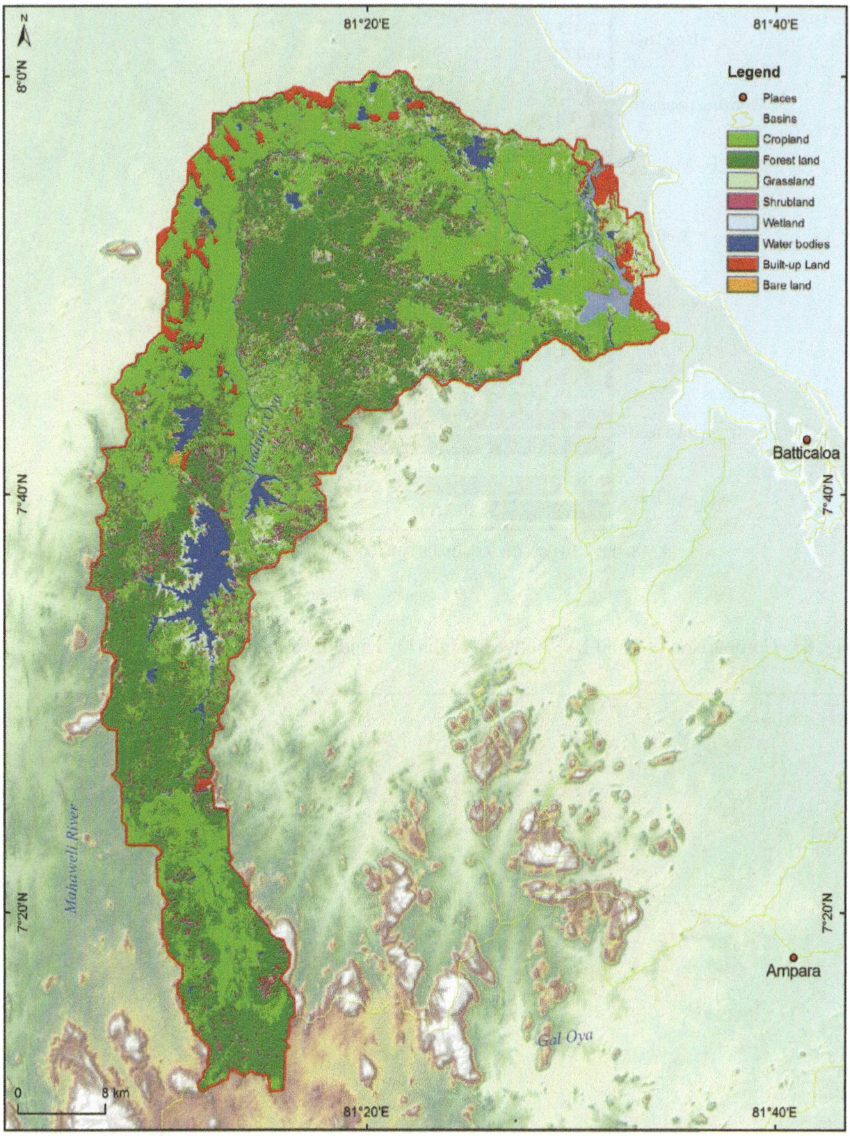

Fig. 83 Land classification of Maduru Oya by remote sensing

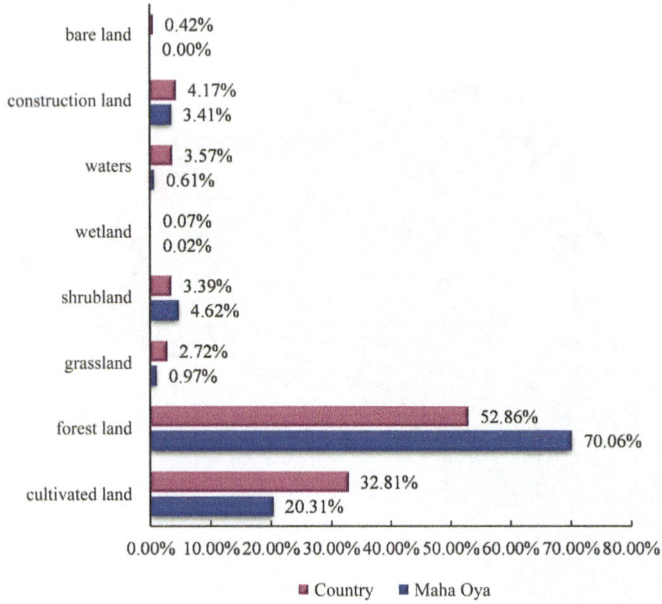

Fig. 84 Comparison of Land Use Between Maha Oya and the country

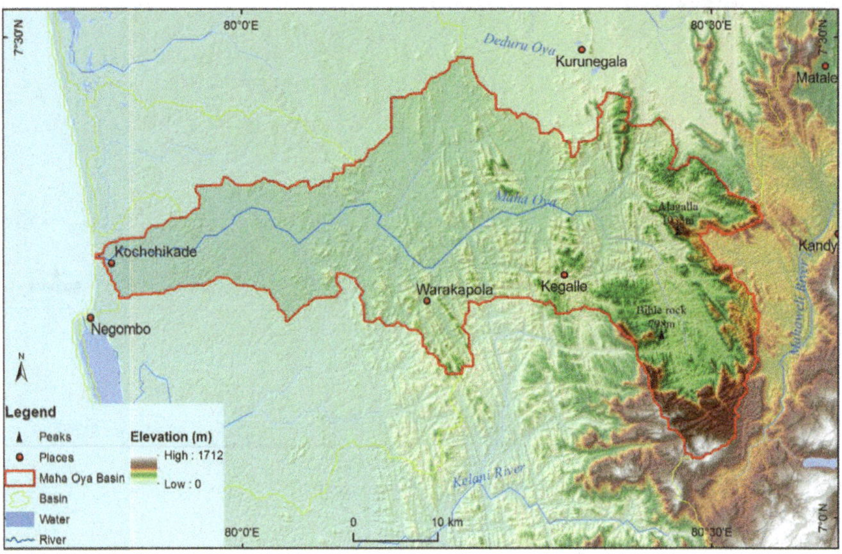

Fig. 85 Topography of Maha Oya

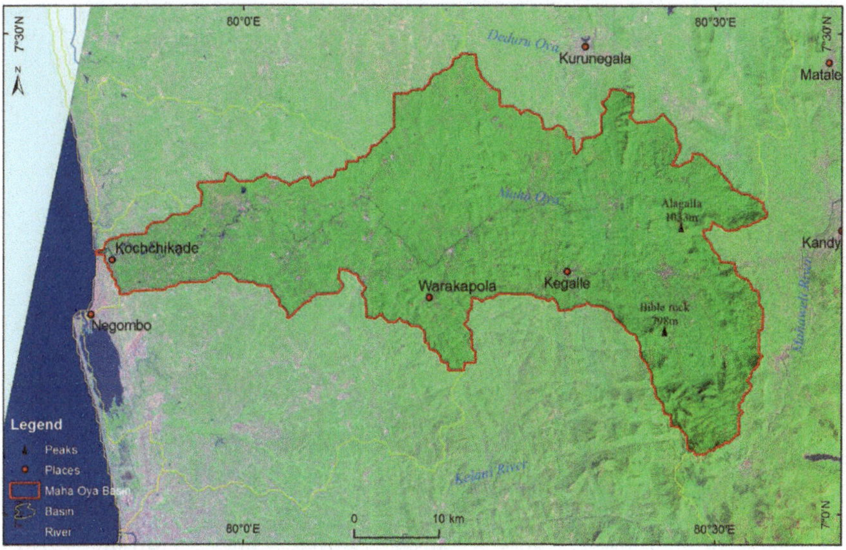

Fig. 86 True color remote sensing image of Maha Oya

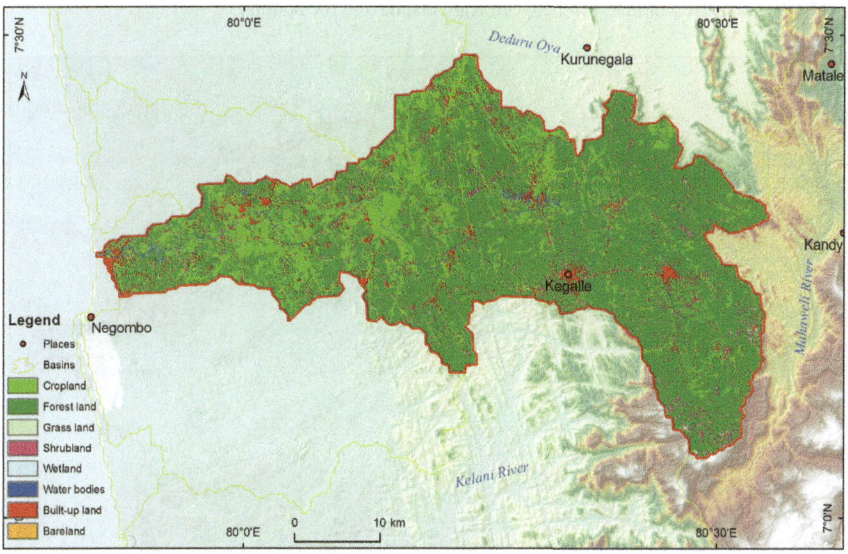

Fig. 87 Land classification of Maha Oya by remote sensing

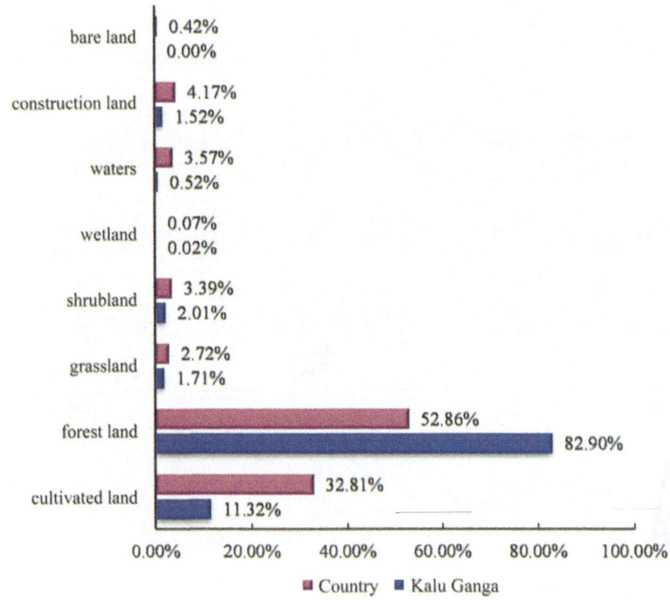

Fig. 88 Comparison of land use between Kalu Ganha and the country

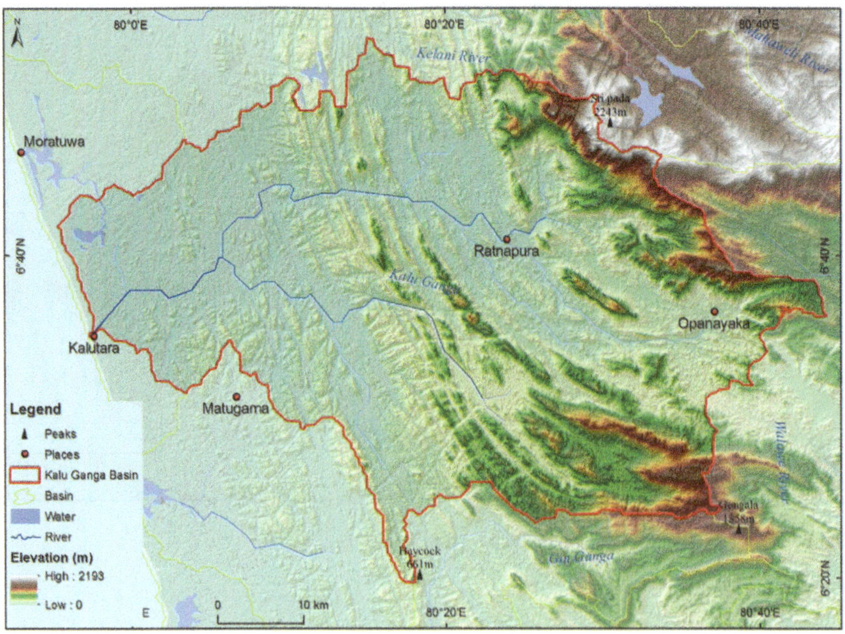

Fig. 89 Topography of Kalu Ganha

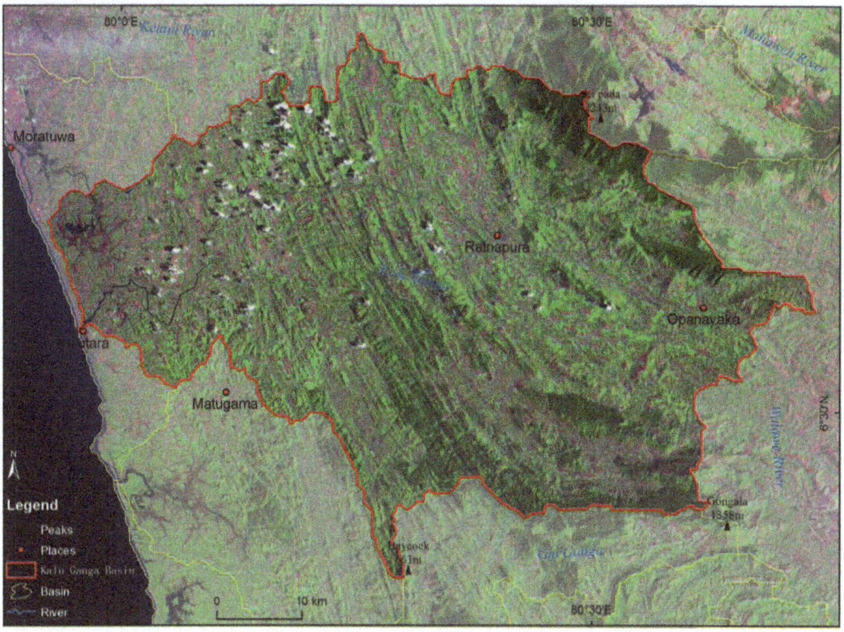

Fig. 90 True color remote sensing image of Kalu Ganha

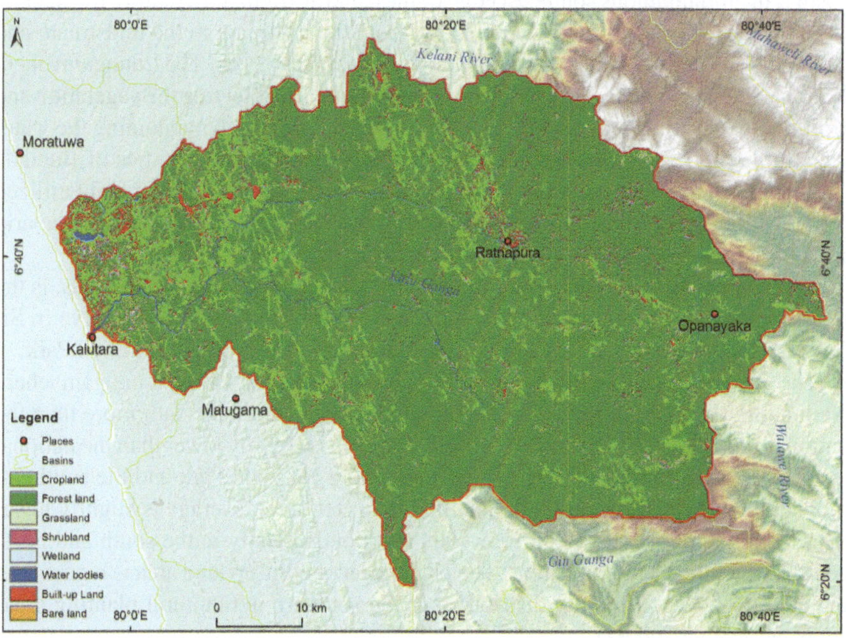

Fig. 91 Land classification of Kalu Ganha by remote sensing

3.11 *Kirindi Oya*

Around 117 km long, it originates from Bandarawela in the southeast of the Central Highlands, which is the second largest town in Badulla District with an altitude of about 1200 m and an annual average rainfall of between 1100 and 1400 mm. The climate is mild throughout the year, making it a tourist attraction for local people. It flows from north to south into the Indian Ocean near the Bundala National Park.

The river locates in the southeast between 6° 12′ N–6° 53′ N and 80° 58′ E–81° 21 ′E. The terrain is high in the north and low in the south peaking at 1,951 m above sea level and average altitude of 236.8 m. The total length is about 80 km and the width ranges from 20 to 30 km, with a total area of 1178 km^2. The north of the basin is the mountainous southeast edge of the Central Highlands, mainly forest land. The altitude drops sharply to below 200 m from the upstream to the midstream and the topography of the midstream to the downstream fairly even. The Lunugamvehera National Park established in 1995 plays a crucial role in protecting the vegetation and wildlife in the Lunugamvehera Reservoir catchment area, and maintaining the water level of the five reservoirs downstream and the wetland characteristics of Bundala National Park near the estuary. The south of the basin is a flat plain, mainly utilized for irrigation agriculture. Most of basin along the Kirindi Oya is located in the arid area with a relatively low annual rainfall of 1485 mm.

Bangdala National Park, an important wintering place for migratory birds, is the first wetland declared as Ramsar Wetland and the fourth Biosphere Reserve in Sri Lanka. There are five shallow salt lakes and three salt fields in the National Park.

The cropland area mainly in the south, north and west of the Lunugamvehera National Park accounts for 36.20% of the total basin area, one tenth more than the national average. The proportion of forest land area is slightly lower than the national average, mainly distributed in the Lunugamvehera National Park and the southeast mountain edge of the Central Highlands. The water area coverage is slightly lower than the national average, with reservoirs and ponds mostly in the south and north central agricultural planting areas. The proportion of built-up land area is higher than the national average, mainly distributed in the southern agricultural planting areas (Figs. 92, 93, 94 and 95).

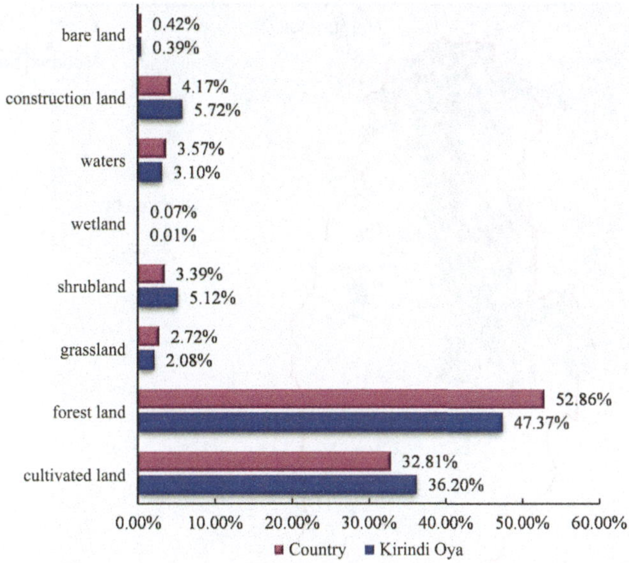

Fig. 92 Comparison of land use between Kirindi Oya and the Country

3.12 *Kumbukkan Oya*

Around 116 km long, the river originates from the southeast edge of the Central Highlands and flows into the ocean near Kumana. It is located in the southeast between 6° 30′ N–7° 05′ N and 81° 10′ E–81° 43′ E. The terrain of the basin is high in the northwest and low in the southeast peaking at 1505 m above sea level and average altitude 201.9 m. The total length of the basin is about 85 km and the width is about 20 km, with a total area of 1233 km^2.

The northwest of the basin is the mountainous southeast edge of the Central Highlands, mainly forest land. The altitude drops sharply to below 200 m from the upstream to the midstream and the topography of the midstream to downstream is fairly flat with elevation below 100 m. Yala National Park established in 1938 in the southwest is famous for a variety of wild animals, which is of importance for

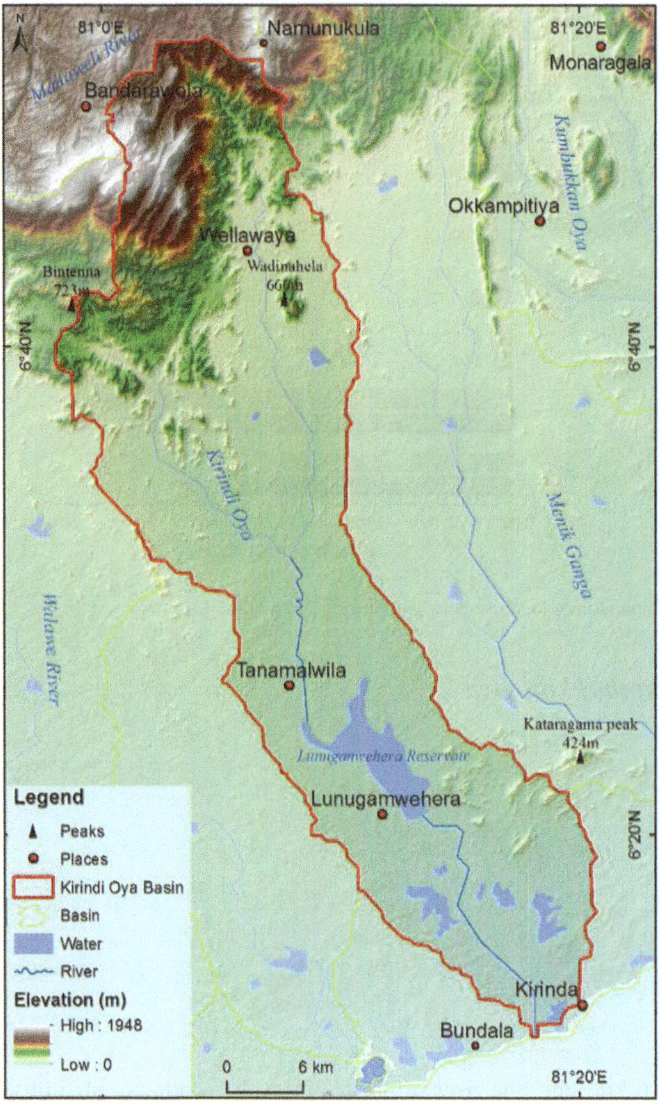

Fig. 93 Topography of Kirindi Oya

the protection of Sri Lanka's elephants and waterfowls. Most of basin along the Kumbukkan Oya is in the arid area, and with a relatively low annual rainfall of 1567 mm.

The cropland area in middle reaches is 21.65% of the total basin area or one thirds of the national average. The forest land area occupies 73.86% of the whole basin, chiefly in the upper reaches and Yala National Park in the middle and low

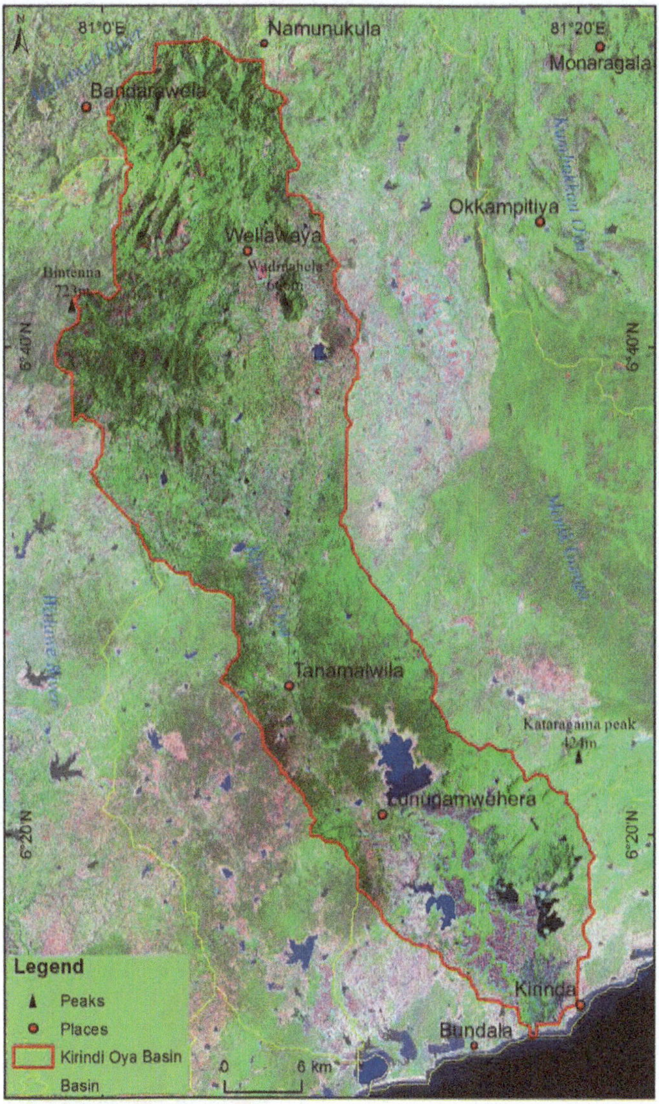

Fig. 94 True color remote sensing image of Kirindi Oya

reaches. The basin has few reservoirs and ponds with the water area coverage of about 0.22%, far lower than the national average. Also, the built-up land area is of quite low proportion (Figs. 96, 97, 98 and 99).

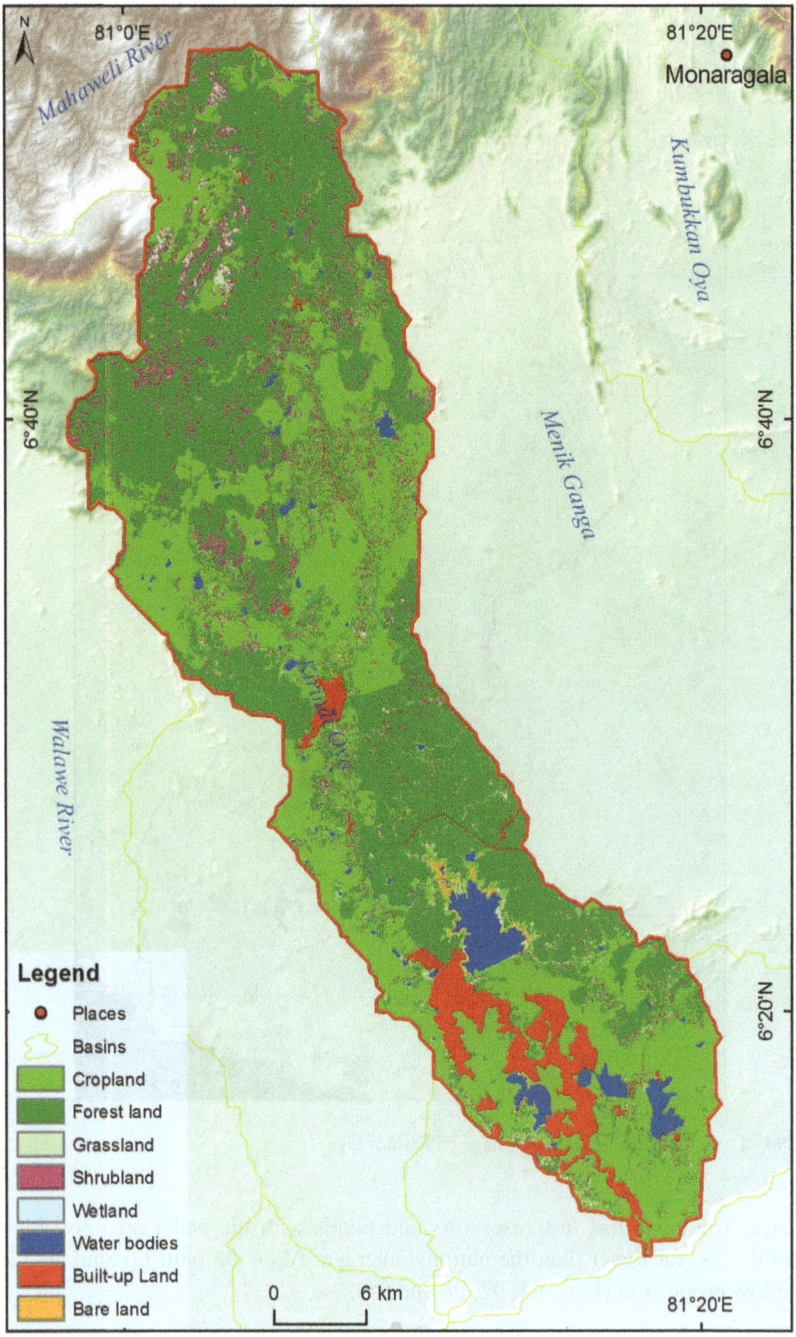

Fig. 95 Land Classification of Kirindi Oya by remote sensing

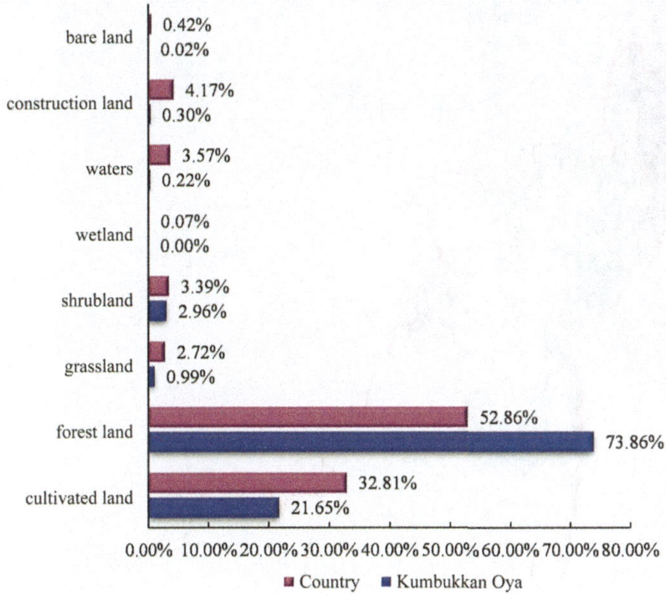

Fig. 96 Comparison of land use between Kumbukkan Oya and the country

3.13 Menik Ganga

Around 114 km long, it originates from Namunukula Mountain in the southeast of the Central Highlands, flows from north to south and turns to east in Kataragama into the ocean near Yala.

The river locates in the southeast between 6° 22′ N–6° 57′ N and 81° 6′ E–81° 33′ E. The terrain is high in the north and low in the south peaking at 1723 m above sea level and an average altitude of 170.0 m. The total length is about 77 km and the width of upper and lower reaches is about 10 km and 30 km respectively, with a total area of 1287 km^2. The north end is the mountainous southeast edge of the Central Highlands, mostly forest land. The south elevation drops rapidly to below

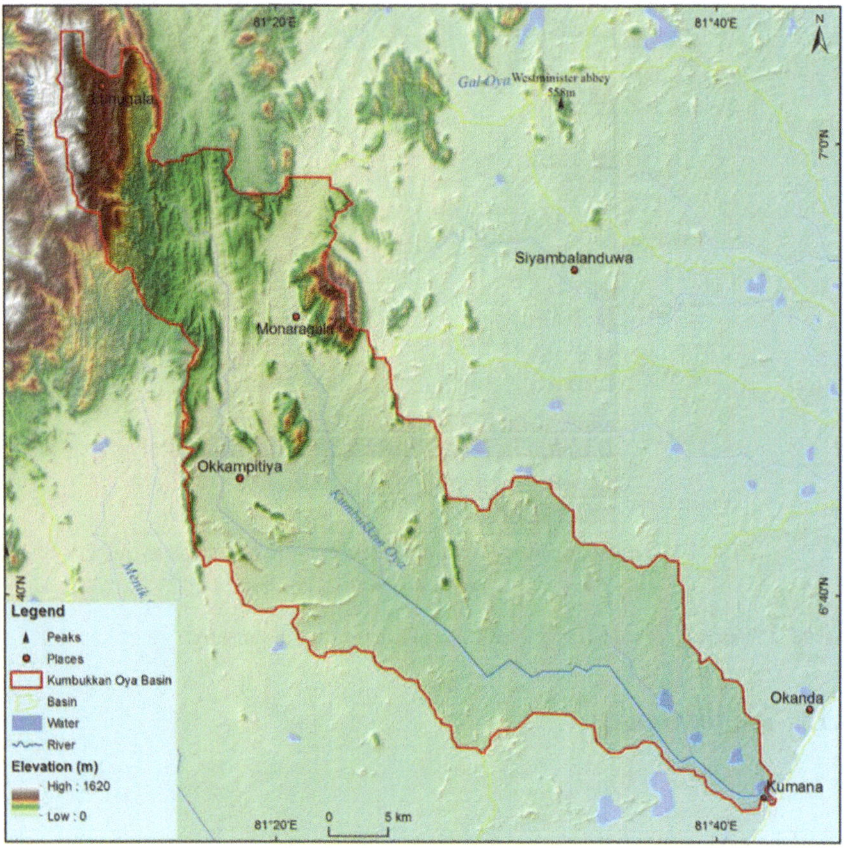

Fig. 97 Topographic of Kumbukkan Oya

200 m. The middle and lower reaches are relatively flat, more than half of which is occupied by Yala National Park.

The cropland area mainly in the middle reaches and Yala National Park accounts for 23.62% of the total basin area or three quarters of the national average. The proportion of forest land area is significantly large, mainly distributed in the National Park in the middle and lower reaches, and on the southeast edge of the Central Highlands. The water area coverage is slightly lower than the national average, with reservoirs and ponds mostly in agricultural planting areas. Further, the proportion of built-up land area is quite low as compared to other basins (Figs. 100, 101, 102 and 103).

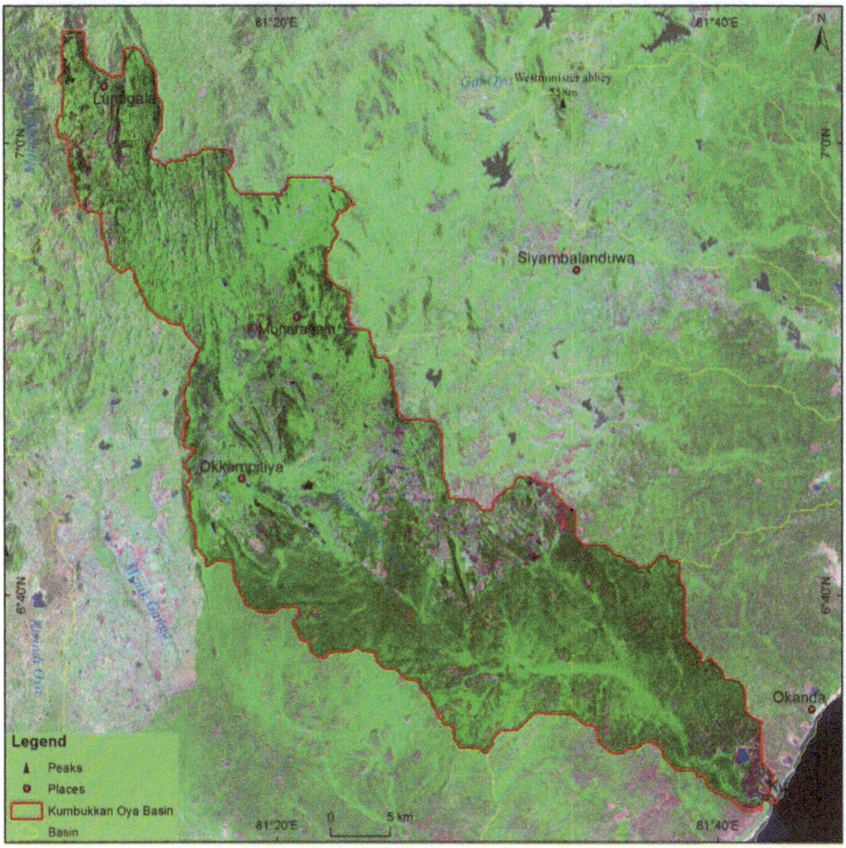

Fig. 98 True color remote sensing image of Kumbukkan Oya

3.14 Gin Ganga

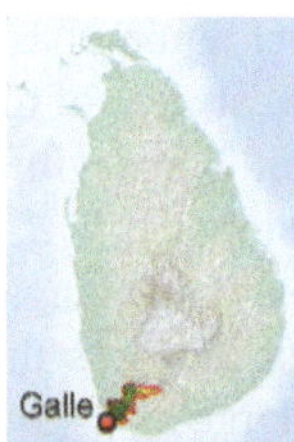

About 113 km long, it originates from Sinharaja Forest Reserve and flows into the ocean near Galle from northeast to southwest.

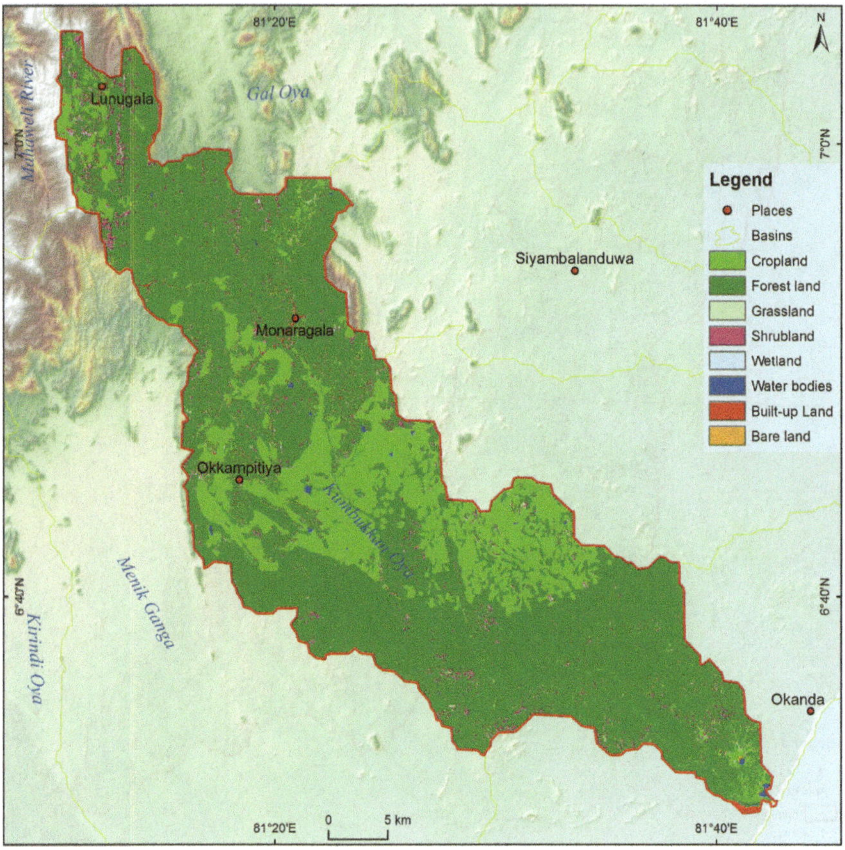

Fig. 99 Land classification of Kumbukkan Oya by remote sensing

It is located in the southwest between 6° 03′ N–6° 26′ N and 80° 08′ E–80° 39′ E. The terrain is high in the northeast and low in the southwest peaking at 1344 m above sea level and average altitude 189.6 m. The total length is about 65 km and the width of upper and lower reaches is about 10 km and 25 km respectively and a total area of 932 km^2, making it the smallest river among basins longer than 100 km. The upper reaches are the mountainous southwest edge of the Central Highlands, mainly forest land, while the middle reaches are hilly area with an altitude of about 200 m, and the lower reaches are relatively gentle slopes.

The proportion of cropland area in Gin Ganga river basin, mainly in strips along rivers or valleys, is relatively small and less than half the national average. The proportion of forest land area is as large as 1.45 times of the national average. With average rainfall of 3285 mm, ranking second across the country, there are very few reservoirs and ponds and thus small water area coverage which is slightly lower than the national average. The proportion of built-up land area mainly in the southwest coast is slightly lower than the national average (Figs. 104, 105, 106 and 107).

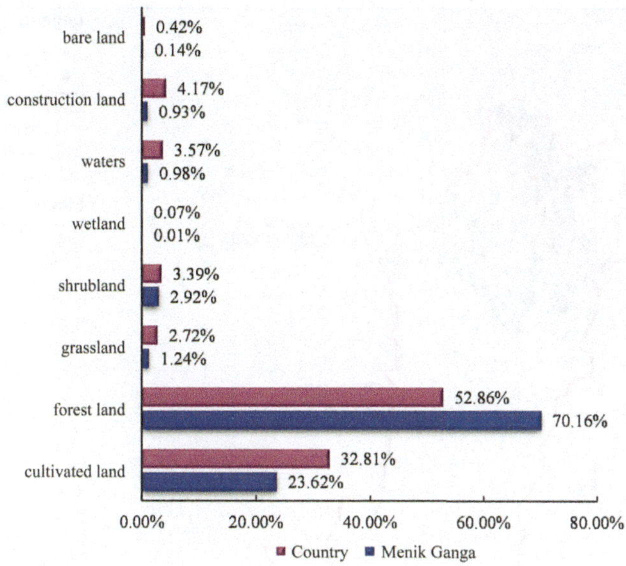

Fig. 100 Comparison of land use between Menik Ganga and the country

3.15 *Mi Oya*

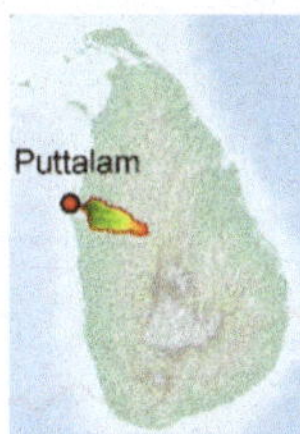

Around 109 km, it originates from Saliyagama and flows into the ocean near Puttalam from east to west.

Mi Oya is located in the northwest between 7° 49′ N–8° 10′ N and 79° 49′ E–80° 29′ E. The terrain is flat except few hills in the upper reaches peaking at 565 m above sea level and average altitude of 76.0 m. The total length is about 78 km and the average width is about 10–25 km, with a total area of 1533 km^2.

The proportion of cropland area is slightly higher than the national average, mainly irrigation agriculture distributed along both sides of the river. The area of forest land mainly in the forest reserves in the north of the middle and lower reaches, is slightly larger than the national average. The average rainfall is 995 mm collected for irrigation in many reservoirs and ponds. The proportion of built-up land area is half the national average (Figs. 108, 109, 110 and 111).

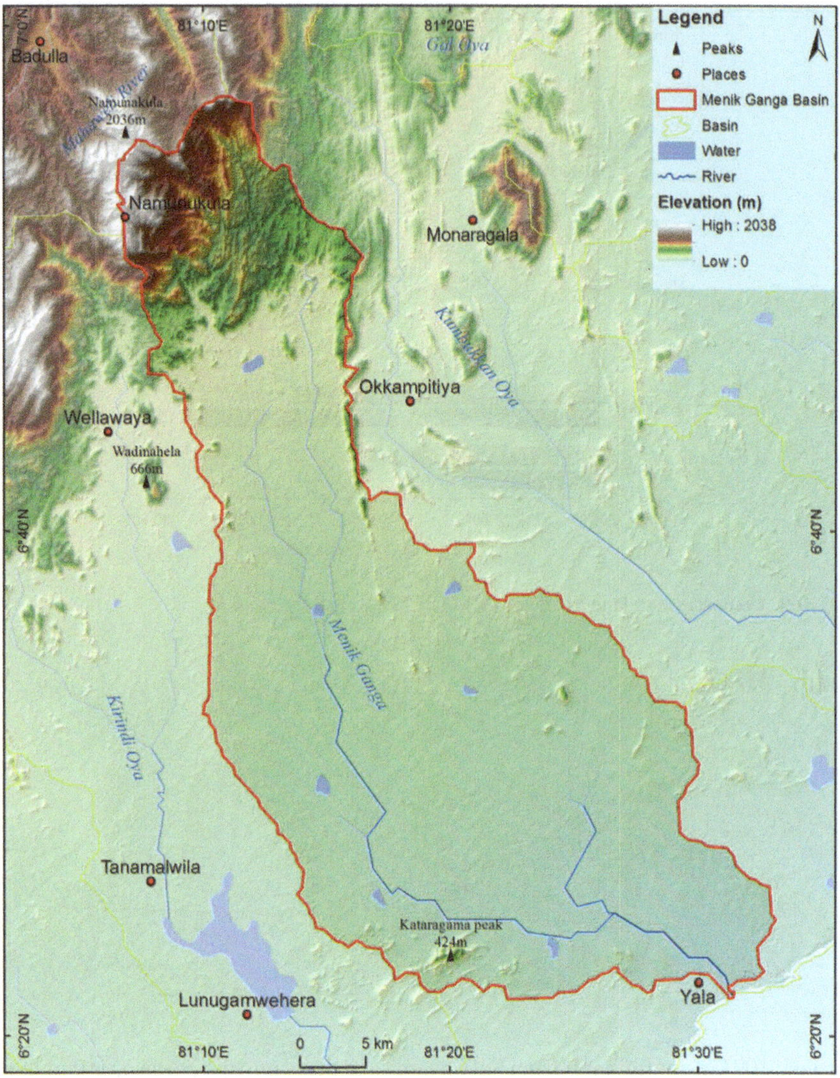

Fig. 101 Topography of Menik Ganga

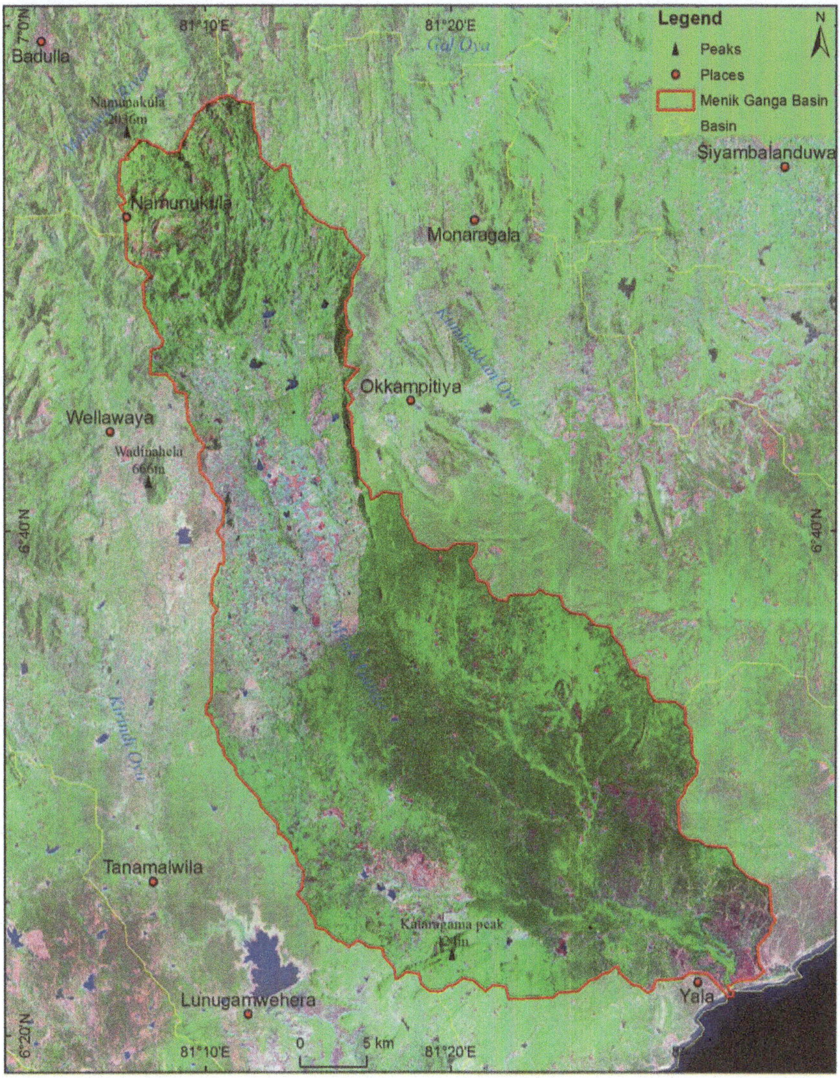

Fig. 102 True color remote sensing image of Menik Ganga

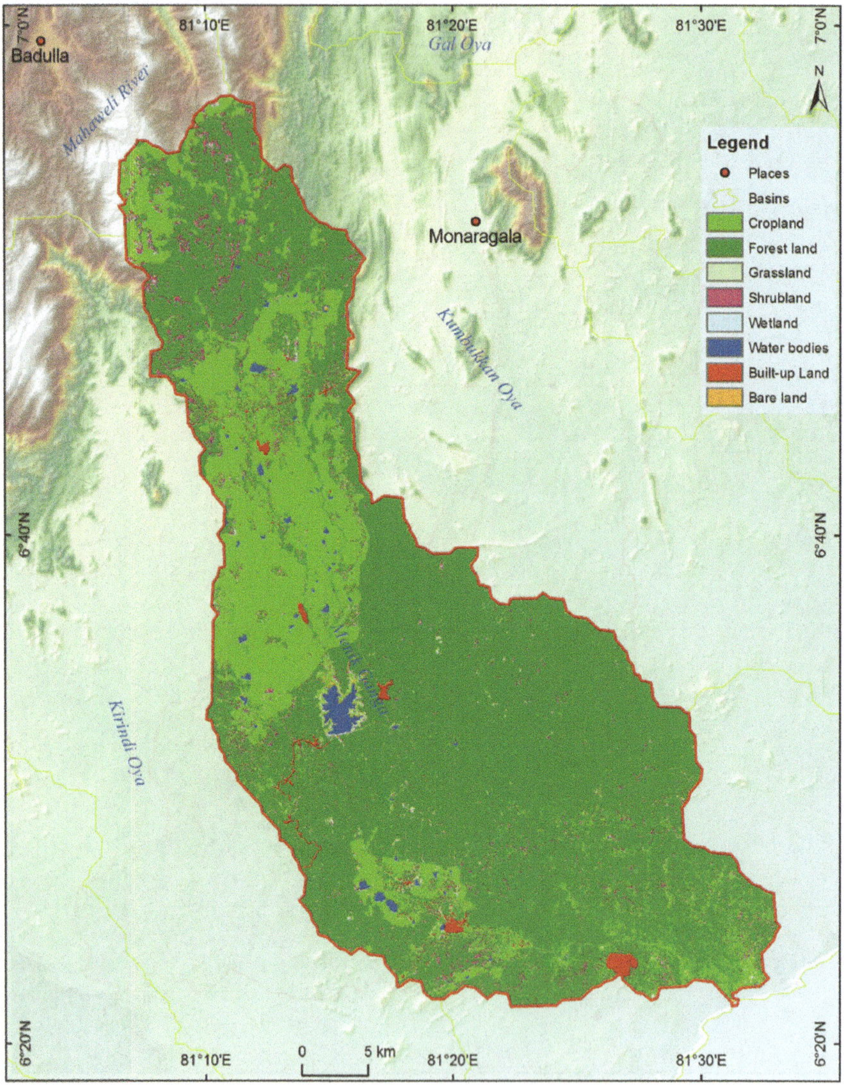

Fig. 103 Land classification of Menik Ganga by remote sensing

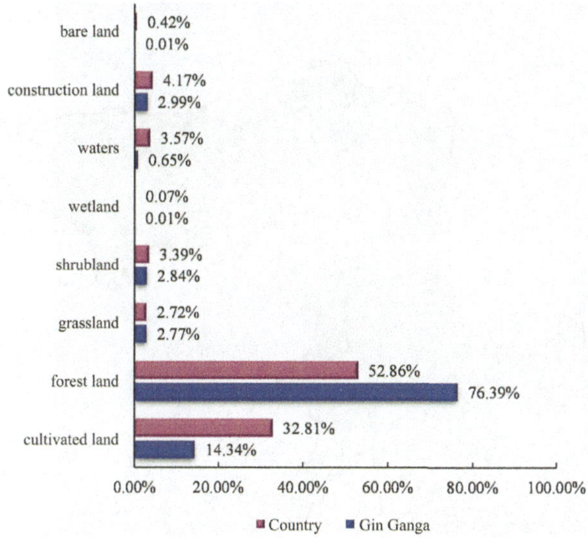

Fig. 104 Comparison of land use between Gin Ganga and the country

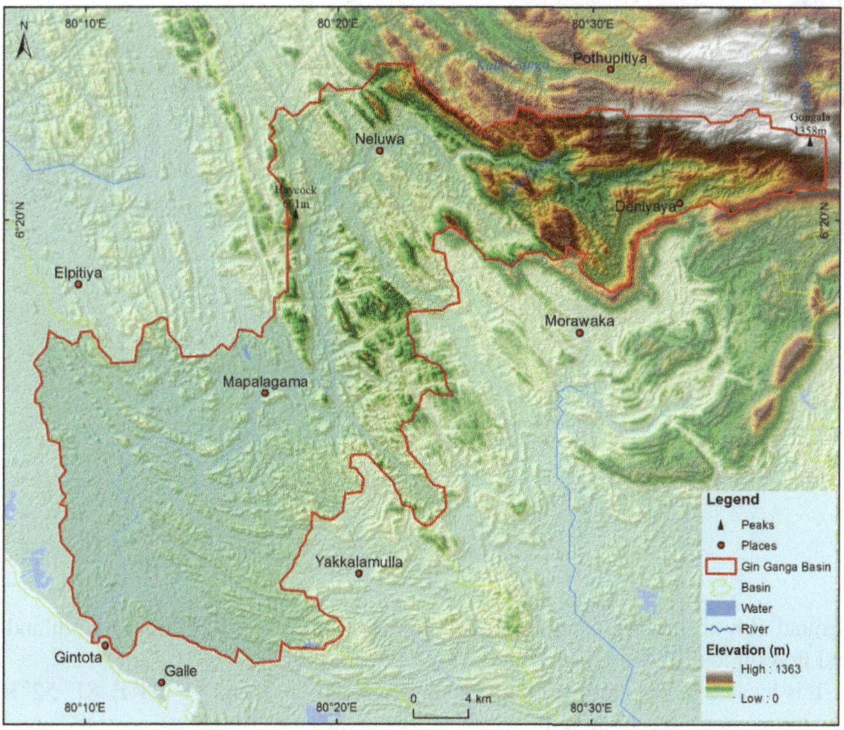

Fig. 105 Topography of figure Gin Ganga

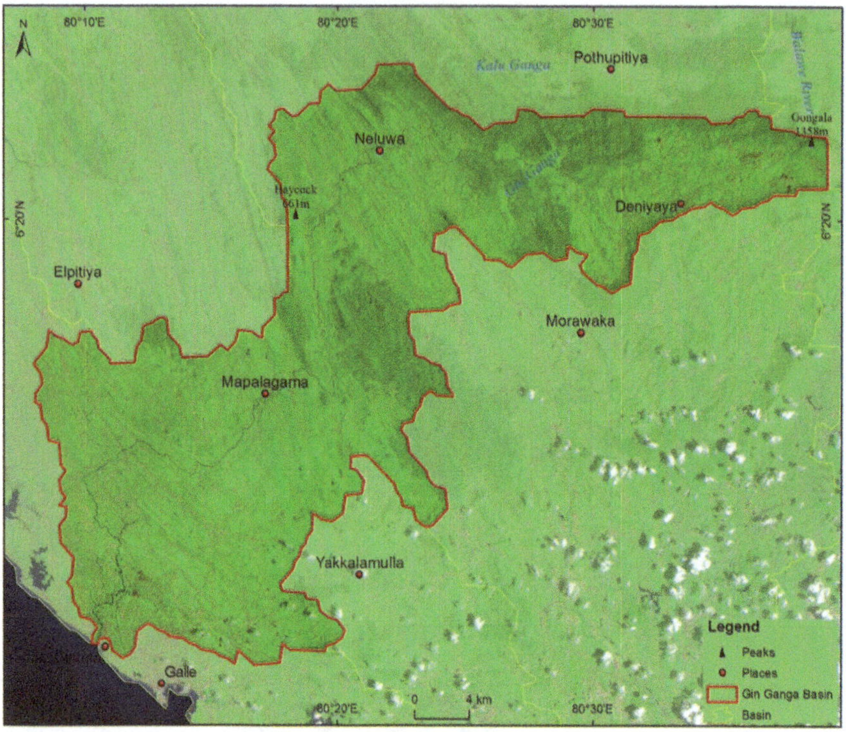

Fig. 106 True color remote sensing image of Gin Ganga

3.16 Gal Oya

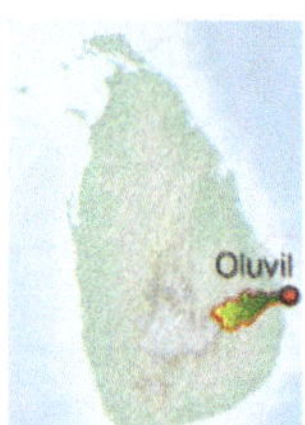

Around 108 km, it originates from Yalkumbura, the east of the Central Highlands and flows into the ocean near Oluvil.

It is located in the southeast between 6° 58′ N–7° 21′ N and 81° 08′ E–81° 52′ E. The basin's terrain is high in the west and low in the east. The elevation of the upper reaches is above 200 m with the westernmost above 1000 m and the highest altitude is about 1510 m. The middle reaches are hilly while the lower reaches are gentle.

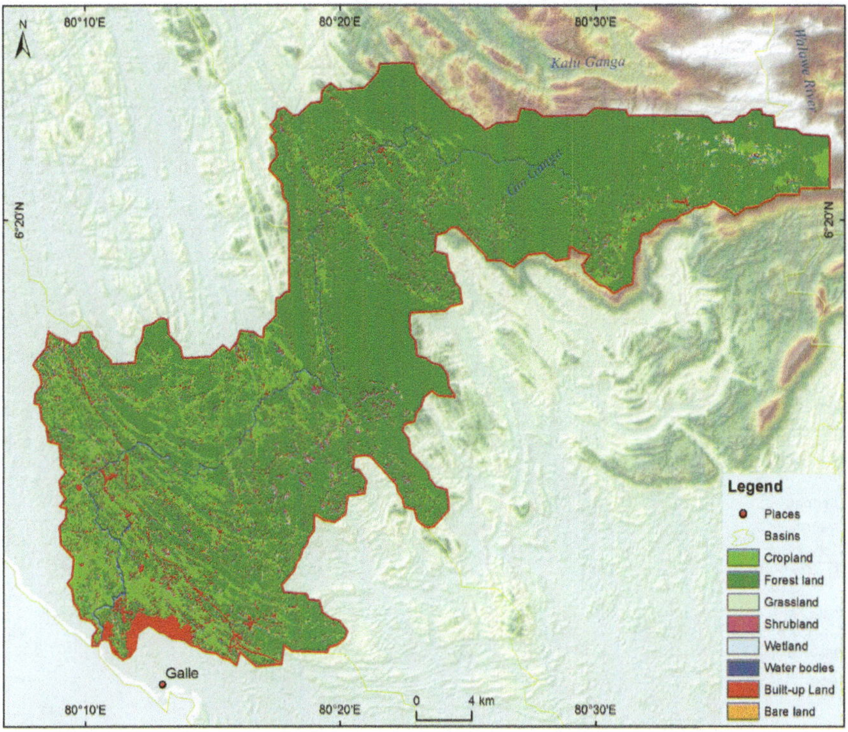

Fig. 107 Land classification of Gin Ganga by remote sensing

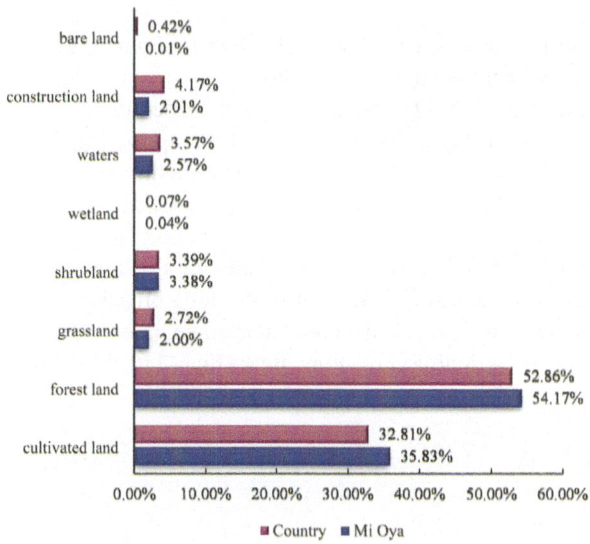

Fig. 108 Comparison of land use between Mi Oya and the country

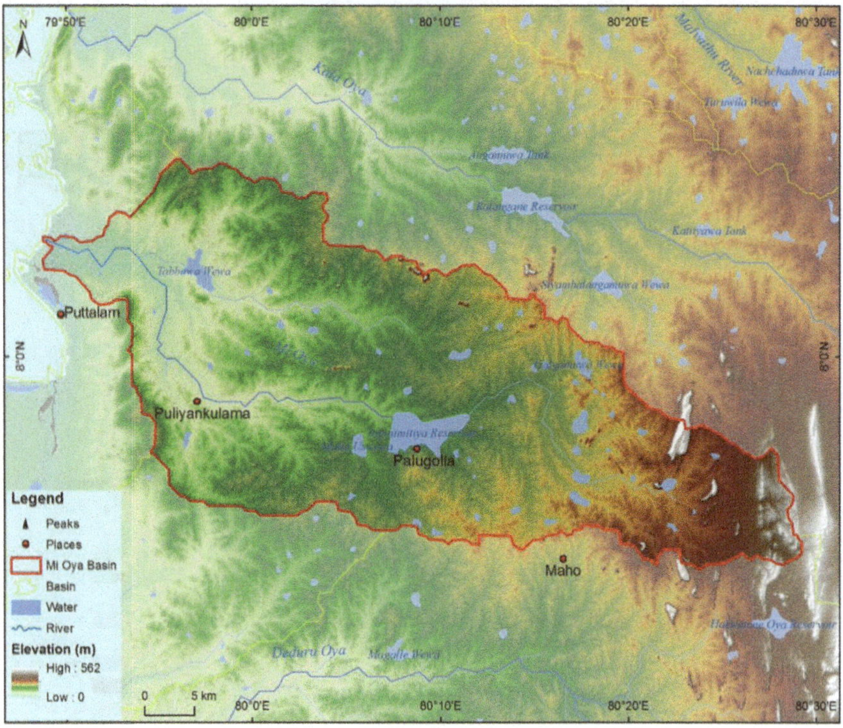

Fig. 109 Topography of Mi Oya

The average elevation is 178.1 m. The total length is about 80 km and the width is about 10–30 km, with a total area of 1813 km^2.

The cropland in the Gal Oya mainly irrigated agriculture, is distributed in the lower reaches, south of the middle reaches, and the upper, which is slightly lower than the national average. The forest land area is slightly larger than the national average, mainly distributed in the middle and upper reaches, especially in the Gal Oya National Park. There are many reservoirs and ponds used for irrigation in the middle and lower reaches with Senanayake Samudra, the large reservoir, providing sufficient water for agricultural irrigation in the lower reaches. The proportion of built-up land area is less than half the national average, mainly distributed in coastal and lower reaches of agricultural planting areas (Figs. 112, 113, 114 and 115).

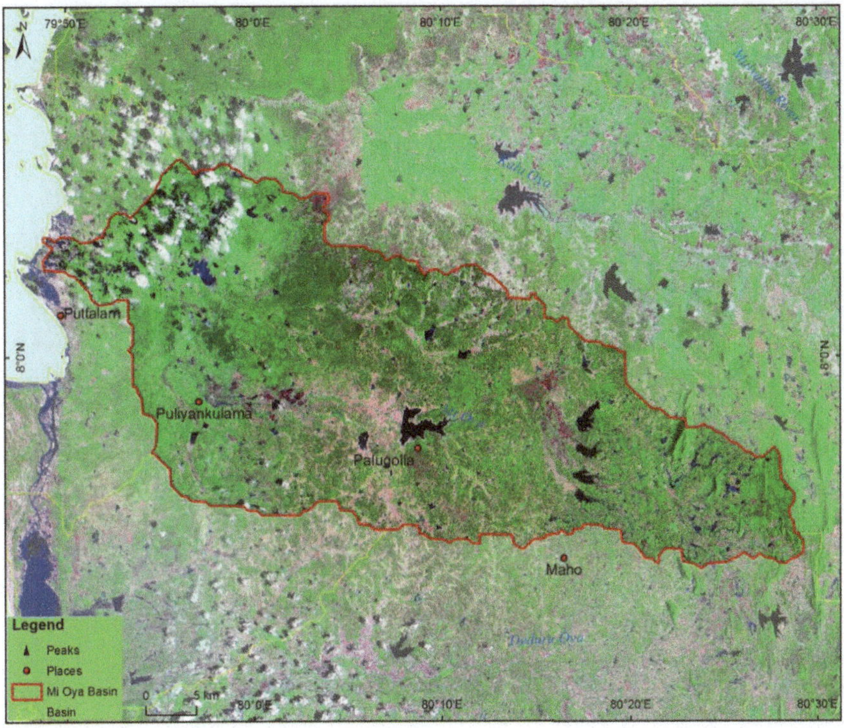

Fig. 110 True color remote sensing image of Mi Oya

4 Lakes and Reservoirs

Lake has certain ecological functions of matter circulation, energy flow and information transmission between lake and terrestrial ecosystem to form the local microclimate and adjust regional climate. The fluctuation of lake volumes is the comprehensive result of water balance in the basin, which is highly sensitive to climate change and human activities and an apparent indicator of climate change. The formation and disappearance, expansion and contraction of lakes, and the changes of ecological environment all reflect regional and or global tectonic and climatic events. Therefore, it is of great significance to monitor the lake's dynamics accurately and quickly, and to reveal the influence of natural factors and human activities on the lake, so as to reasonably develop, utilize and protect the water resources.

Sri Lanka has a tropical monsoon climate with most land in the country is dry zone or intermediate zone. Only part of the southwest has wet zone due to frontal rain caused by the mountains in the middle blocking the warm and humid southwest monsoon. There are also dry seasons and rainy seasons in the arid area. The spatial-temporal distribution of water resources in the island is very uneven and seasonal water shortage is very severe. Consequently, the country has built many reservoirs

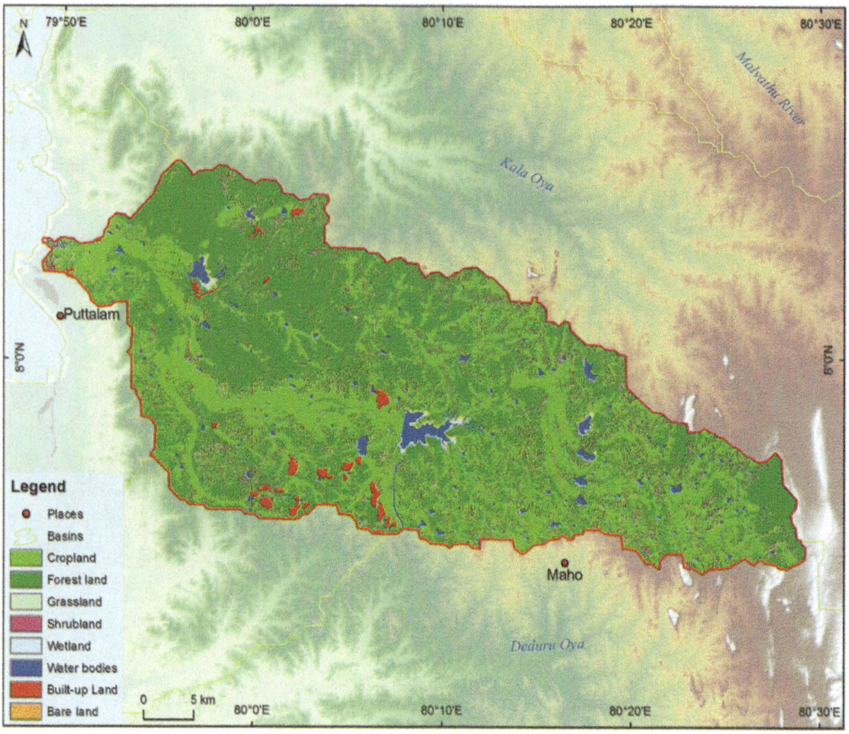

Fig. 111 Land classification of Mi Oya by remote sensing

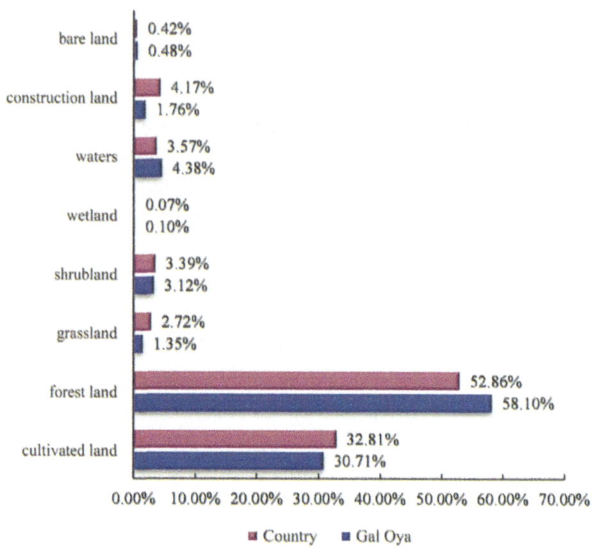

Fig. 112 Comparison of land use between Gal Oya and the country

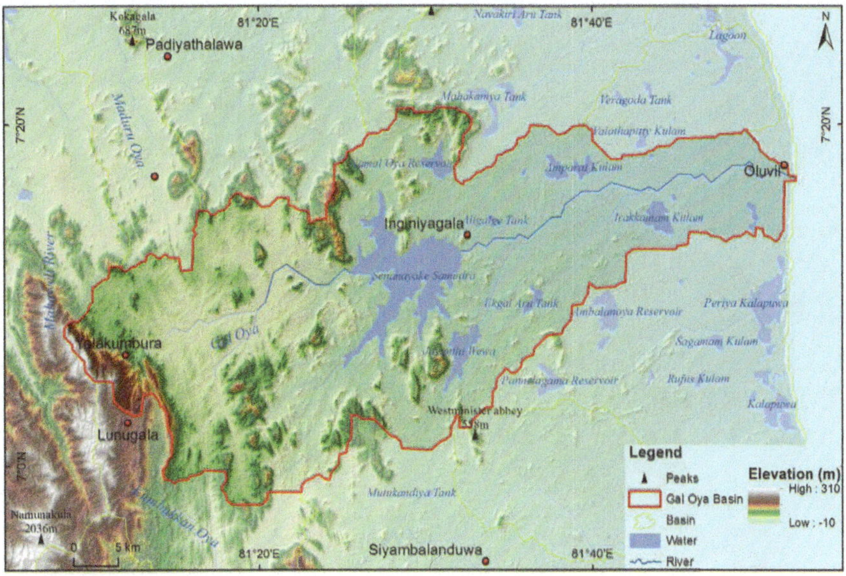

Fig. 113 Topography of Gal Oya

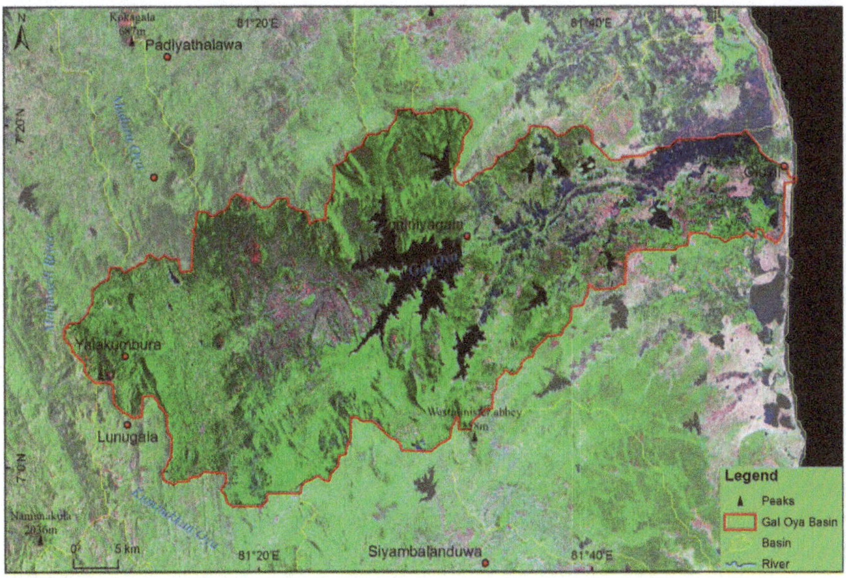

Fig. 114 True color remote sensing image of Gal Oya

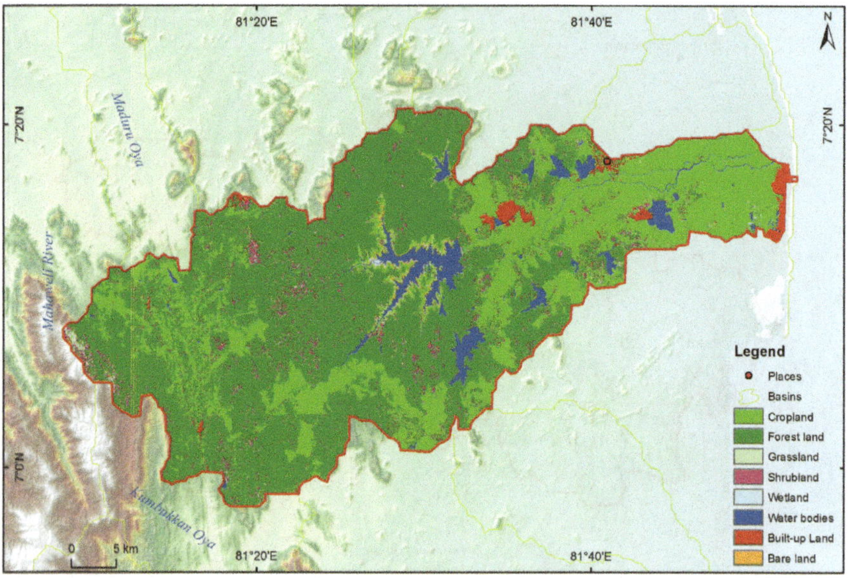

Fig. 115 Land classification of Gal Oya by remote sensing

for agricultural irrigation and domestic water since ancient times. Sri Lanka began developing irrigation since 5 BC and has built more than 1500 water storage pools in the central arid areas especially between the 2 and the 14 AD, forming an irrigation network with many reservoirs, ponds and canals. After the 19th century, it began to rebuild new reservoirs. In recent years, Sri Lanka has developed rapidly in irrigation, flood control, hydroelectric power, domestic and industrial water supply. The growth rate of irrigation area also increased as the number of new large, medium and small-scale irrigation projects have increased significantly. At the same time, the proportion of the pond irrigated area has been considerably grown due to the emphasis on the construction and maintenance of the ponds in the rain fed agricultural areas. The dam network and related water infrastructure are mainly used to meet the needs of farmland irrigation, hydropower, domestic water supply, industrial water supply, and flood control. The country's power generation and production of major food crops are heavily dependent on water resources regulated by dams. Most of the farming population live on the irrigation of farmland in the arid area, which accounts for nearly two thirds of the country's land area. In this section, the dynamic changes of Sri Lanka's lakes and reservoirs are monitored through multi-stage remote sensing images, which can be guidance for flood disaster prediction, monitoring, reduction and assessment, and provides scientific data for water resource protection, management and planning (Fig. 116).

The spatial-temporal characteristics of waters in Sri Lanka in 1995, 2005 and 2015 are analyzed as followed based on the Landsat series remote sensing images, three phases of spatial distribution of national lakes and reservoirs from OTSU and

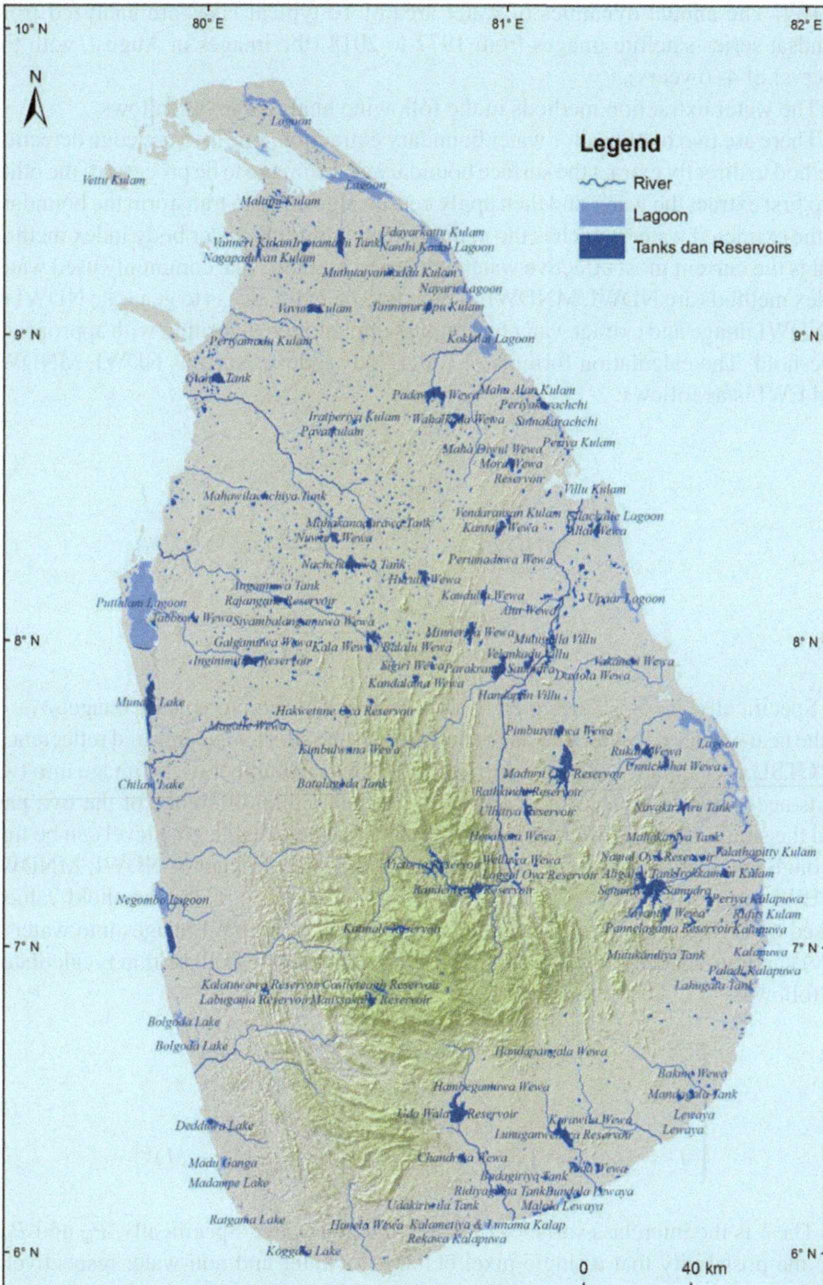

Fig. 116 Distribution of rivers, lakes, reservoirs and ponds

NDWI. The annual dynamics of water area of 10 typical lakes are analyzed from Landsat series satellite images from 1977 to 2018 (the images in August, with the interval of 4–6 years).

The water extraction methods in the following analysis are as follows.

There are two methods for water boundary extraction: one is to use edge detection method to directly extract the surface boundary of the image to be processed; the other is to first extract the water and then apply certain algorithm to transform the boundary of the extracted water, which is the main method including water body index method that is the current most effective water extraction method. The commonly used water index methods are NDWI, MNDWI, EWI, etc. The main idea is to generate NDWI or MNDWI image and extract water by simple segmentation algorithm with appropriate threshold. The calculation formula of water body indexes such as NDWI, MNDWI and EWI is as follows:

$$\mathrm{NDWI} = \frac{\rho_{Green} - \rho_{NIR}}{\rho_{Green} + \rho_{NIR}} \tag{1}$$

$$\mathrm{MNDWI} = \frac{\rho_{Green} - \rho_{SWIR}}{\rho_{Green} + \rho_{SWIR}} \tag{2}$$

$$\mathrm{EWI} = \frac{\rho_{Green} - \rho_{NIR} - \rho_{SWIR}}{\rho_{Green} + \rho_{NIR} + \rho_{SWIR}} \tag{3}$$

Specifically, ρ_{Green} is the green band reflectance of remote sensing image, ρ_{SWIR} is the near-infrared band reflectance and ρ_{SWIR} is the short-wave infrared reflectance.

OTSU uses the idea of clustering to divide the gray number of the image into two parts according to their level to ensure the largest value difference of the two part and the smallest value difference inside each part. The suitable gray level can be find through calculation of variance. Suppose that the values range of NDWI, MNDWI or EWI images is x to y, with the range of $-1 \leq \mathrm{x} < y \leq 1$, the threshold value t based on OTSU algorithm can divide NDWI, MNDWI or EWI images into water (t to y) and non-water (x to t). The optimal threshold t of OTSU algorithm is calculated as follows:

$$\begin{cases} \delta^2 = P_{nw} \cdot (M_{nw} - M)^2 + P_w(M_w - M)^2 \\ M = P_{nw} \cdot M_{nw} + P_w \cdot M_w \\ P_{nw} + P_w = 1 \\ t = Arg \operatorname*{Max}_{x \leq t \leq y} \left\{ P_{nw} \cdot (M_{nw} - M)^2 + P_w(M_w - M)^2 \right\} \end{cases} \tag{4}$$

The δ is the interclass variance of water and non-water. Specifically, P_w and P_{nw} are the possibility that a single pixel belongs to water and non-water respectively, M_w and M_{nw} are the gray-scale average value of water and non-water respectively, and M is the gray-scale average value of NDWI, MNDWI or EWI images.

4.1 Spatial Temporal Variations of Lakes and Reservoirs

The remote sensing monitoring area of the national surface water (excluding river area) was 1653.91 km^2, 1964.52 km^2 and 2136.57 km^2 in 1995, 2005 and 2015 respectively. During this period, the national water area increased drastically by 483.66 km^2.

The total water area of the lakes and reservoirs including lagoons is generally increasing. Accounting for inland water excluding lagoons, statistics shows that the total water area was 1020.37 km^2 in 1995, 1270.48 km^2 in 2005 and 1417.48 km^2 in 2015. During two decades, the area of fresh water reservoir has increased by 397.11 km^2 (Fig. 117).

In order to further analyze the spatial-temporal dynamics of inland waters, it can be divided into four grades: Grade I, micro lakes and reservoirs (<0.1 km^2); grade II, small lakes and reservoirs ($\geq$0.1–1 km^2); grade III, medium-sized lakes and reservoirs ($\geq$1–5 km^2); grade IV, large lakes and reservoirs ($\geq$5 km^2). Statistics indicates that the number of lakes and reservoirs in the four grades showed an increasing from 1995 to 2015. The smaller the lakes and reservoirs grades, the larger the number of the lakes and reservoirs, and the greater the number increase of the lakes and reservoirs. The number of micro lakes and reservoirs increased a lot while the number of large lakes and reservoirs increased by 11. Of the new increased large lakes and reservoirs, 7 are transformed from other grade waters and 4 are new built reservoirs, namely Rambukkam Oya, Weheragala, Daduru Oya and Mau Ara. The area of all the grades showed an increasing trend from 1995 to 2015, where the area of large lakes and reservoirs increased by 197.49 km^2, equaling to the total area increase of the other three water grades. In addition, the higher the water grades, the larger the total area of the corresponding grade of lakes and reservoirs. In 2015, the total area

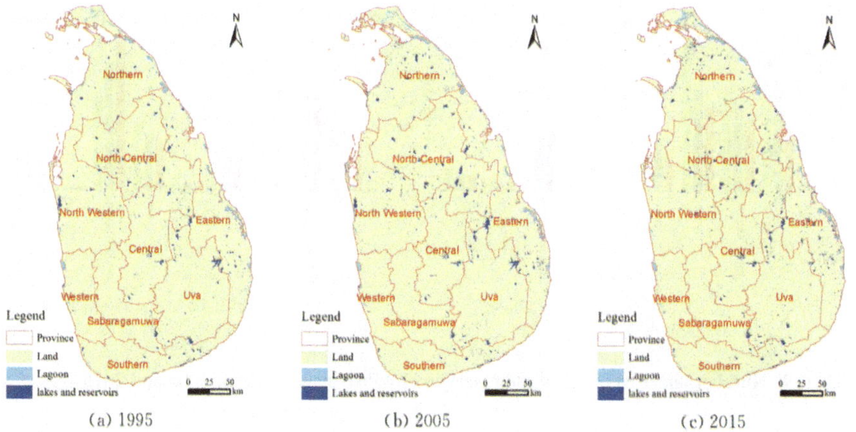

Fig. 117 Lakes, reservoirs and lagoon distribution in 1995, 2005 and 2015

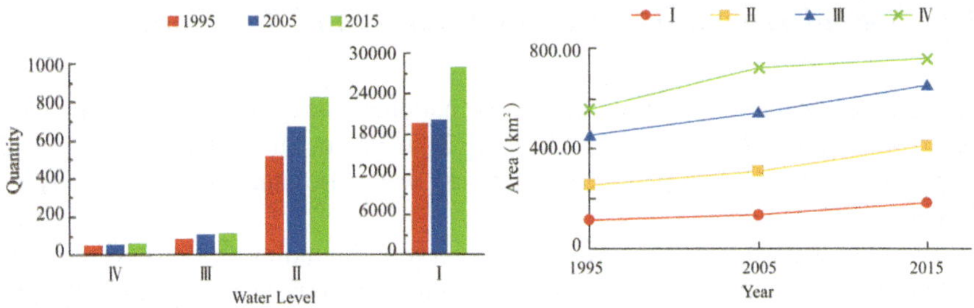

(a) The number of lakes and reservoirs in 1995, 2005 and 2015

(b) The area of lakes and reservoirs in 1995, 2005 and 2015

Fig. 118 Number and area of lakes and reservoirs in 1995, 2005 and 2015

of large lakes and reservoirs reached 760.08 km^2, accounting for 54% of the total area water (Fig. 118).

The number and area of waters changed significantly from 1995 to 2015 and those increased ones were mainly concentrated in arid areas in Northern, North Central, Eastern, Sabaragamuwa, Uwa and Central provinces. Geographically, the number and area of lakes in Southern province were constant, while that of North Western and Western decreased slightly with decreasing number and area mainly corresponded to the lower local rainfall. Contrastingly, the soaring number and area is the result of the increase of local rainfall and the construction of water conservancy facilities. In recent years, Sri Lanka has built a large number of new water conservancy facilities to support agricultural irrigation, aquaculture for local economic development, which can play a role in regulating the water distribution in rainy and dry seasons. In general, the number and area of all lakes increased and the available water resources showed the same trend (Fig. 119).

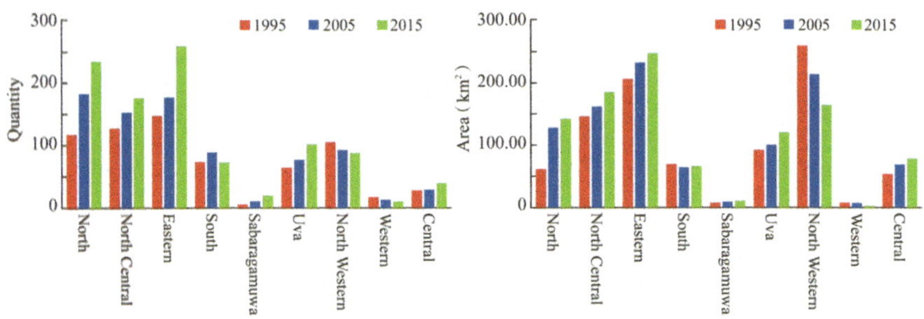

(a) The number of lakes and reservoirs by provinces in 1995, 2005 and 2015

(b) The area oflakes and reservoirs by provinces in 1995, 2005 and 2015

Fig. 119 Number and area of lakes and reservoirs by Provinces in 1995, 2005 and 2015

4.2 Analysis of Spatial-Temporal Area Dynamics of Typical Lake

Senanayake Samudra

Located along the Gal Oya, Senanayake Samudra is one of the largest and most representative reservoirs in the country, also known as Gal Oya Reservoir. The reservoir was formed after the completion of Gal Oya Dam (also known as Inginiyagala Damgaloya dam). Gal Oya is located in Uwa province and was built between two hills in the town of Inginiyagala on August 24, 1949 and completed in 1953. It has a length and a height of 1100 m and 43 m respectively, and a storage capacity of 950 million cubic meters with an area of 91 km^2. The reservoir is used for agricultural irrigation in Uwa and Eastern provinces as well as power generation.

Area dynamics of Senanayake Samudra can be obtained based on the water area characteristic through Landsat remote sensing images from 1977 to 2017. Statistics shows that the area varied a lot in the same period but remained fairly stable from 1991 to 2012. The lake area peaked at 45.41 km^2 in 2006 from the lowest area of about 11.50 km^2 in 1977 (Figs. 120, 121, 122 and 123).

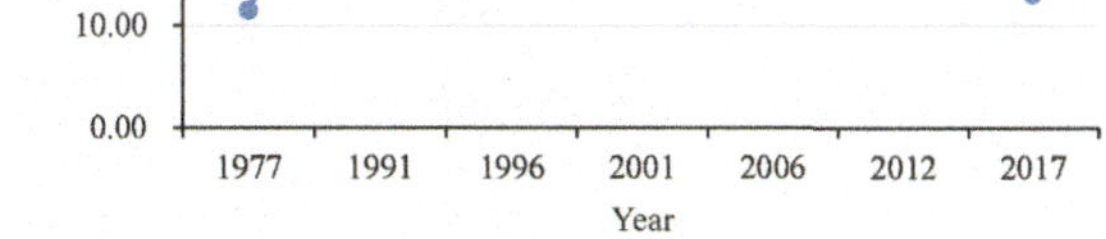

Fig. 120 Area dynamics of Senanayake Samudra

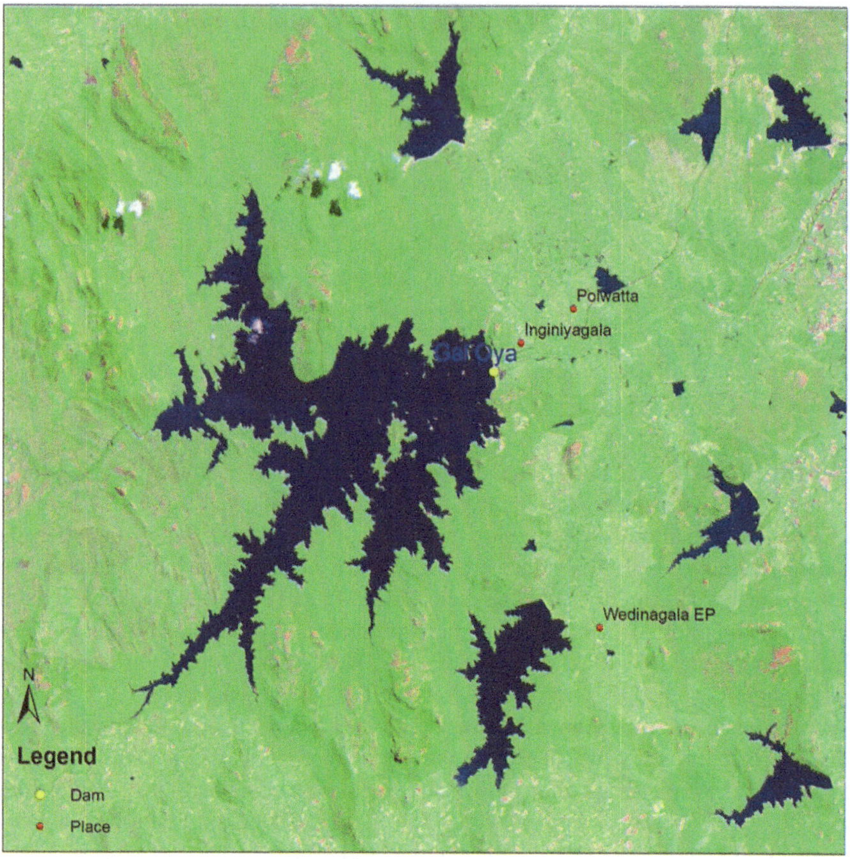

Fig. 121 Remote sensing image of Senanayake Samudra

Maduru Oya Reservoir

Maduru Oya reservoir along Maduru Oya has a catchment area of about 453 km^2 and a reservoir capacity of 5.96 hundred million cubic meters. It is an irrigation dam with a length of 1090 m and a height of 41 m.

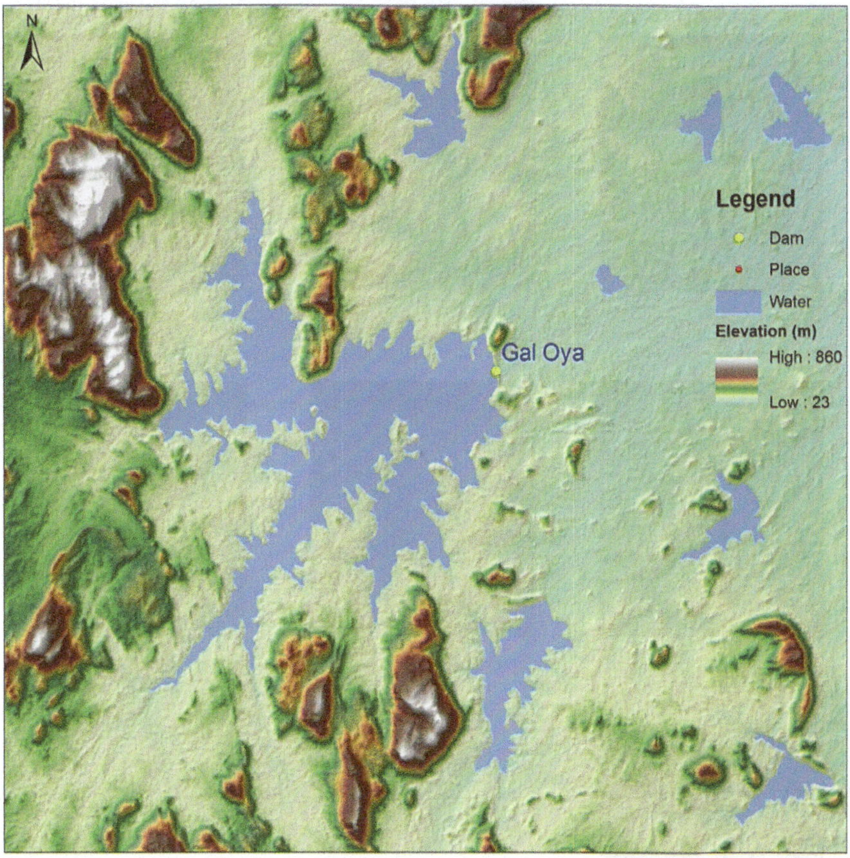

Fig. 122 Topography of the surrounding area of Senanayake Samudra

The water area information can be extracted from Landsat remote sensing images from 1988 to 2018. The results indicates that its area was relatively steady and the surface area remained above 25 km^2 in the past 30 years with the largest surface area of 42.95 km^2 in 2013, and the smallest (27.4 km^2) in 1992. In general, the area shows increasing trend coupled with small fluctuation in the past 30 years (Figs. 124, 125, 126 and 127).

Kantale Tank

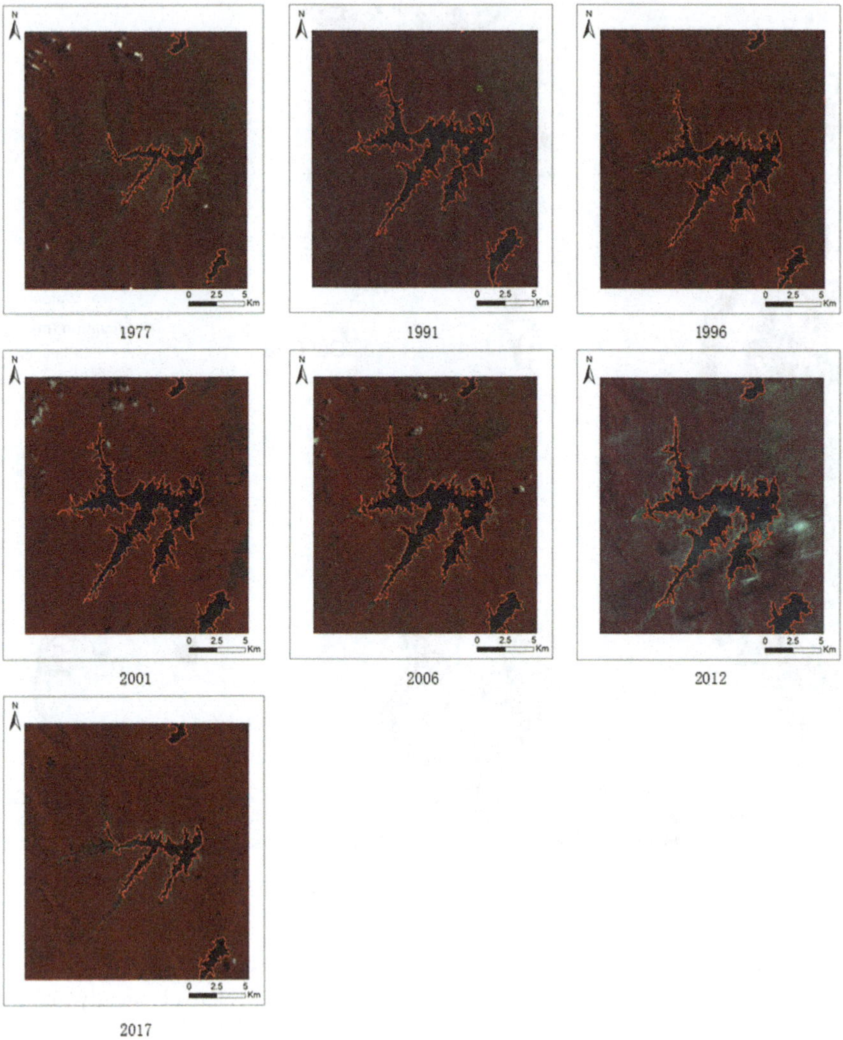

Fig. 123 Remote sensing images of Senanayake Samudra in Different Period

Kantale Tank is a large reservoir on the west of Kantale City utilized for irrigation. Kantale Tank is 4267 m long and 15 m deep. Although having burst on April 20, 1986 and killed more than 120 people, it has been rebuilt.

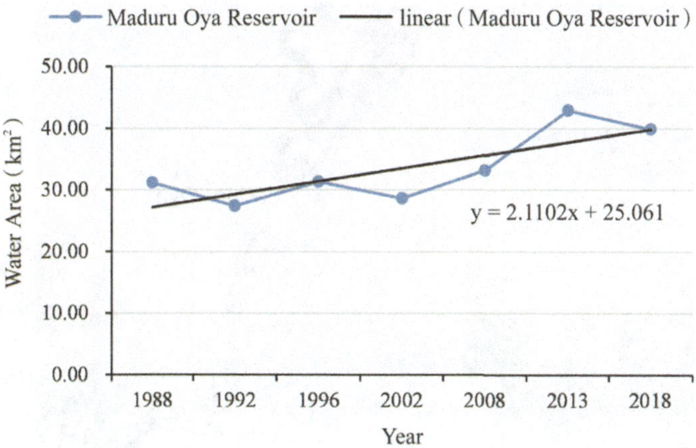

Fig. 124 Area dynamics of Maduru Oya Reservoir

Fig. 125 True color remote sensing image of Maduru Oya Reservoir

Fig. 126 Topography of Maduru Oya Reservoir

The water area information can be extracted from Landsat remote sensing images from 1977 to 2018. That indicates that its surface area in the past 40 years fluctuated significantly peak in at 17.48 km^2 in 2013 and the lowest (5.23 km^2) in 2004 (Figs. 128, 129, 130 and 131).

Uda Walawe Reservoir

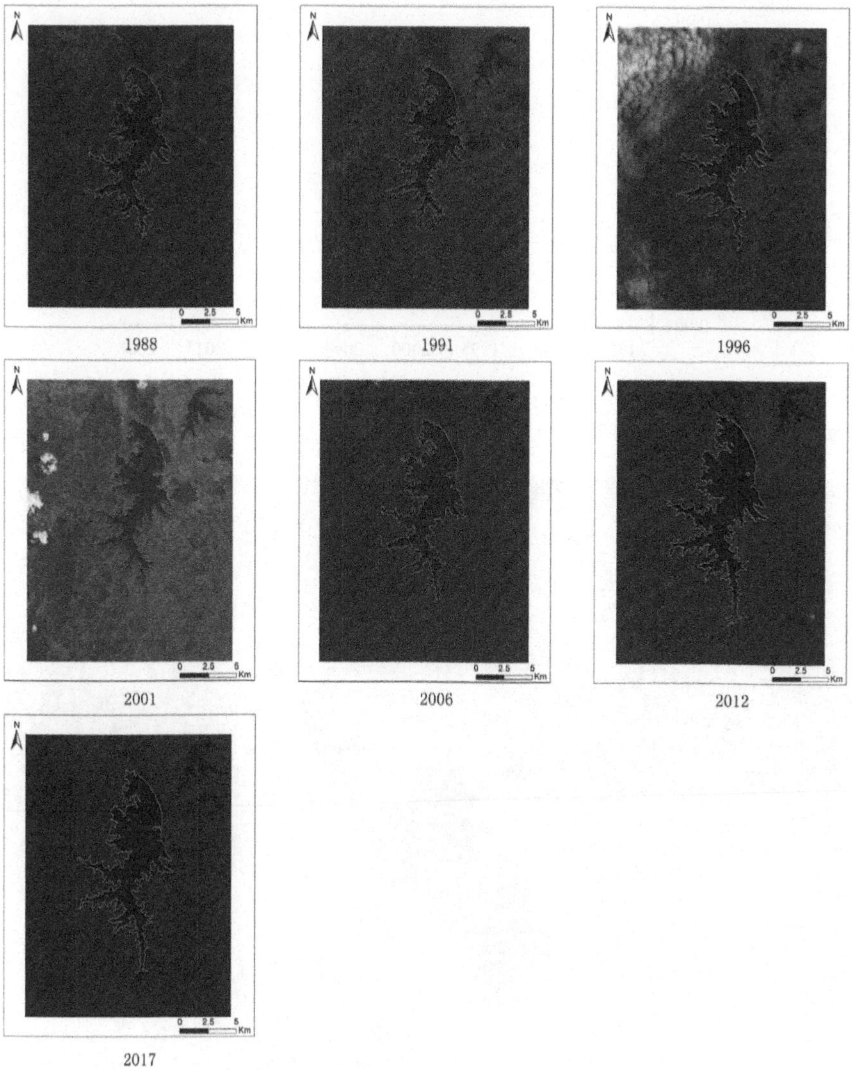

Fig. 127 Remote sensing image of Maduru Oya Reservoir in different periods

Uda Walawe Reservoir, located in the middle of the valley along Uda Walawe, is a large irrigation reservoir on the north of Udawalawa in the south region of Sri Lanka. The water level of the reservoir is controlled by five gate spillways at the eastern end of the dam and other two at the center of the dam. The length of the reservoir is about 8.6 km and the maximum width is about 7 km. The dam is also used for hydroelectric power for three 2-MW units and was put into operation in April 1969.

The water area information can be extracted from Landsat remote sensing images from 1988 to 2018. The result indicates that its surface area has fluctuated in the past

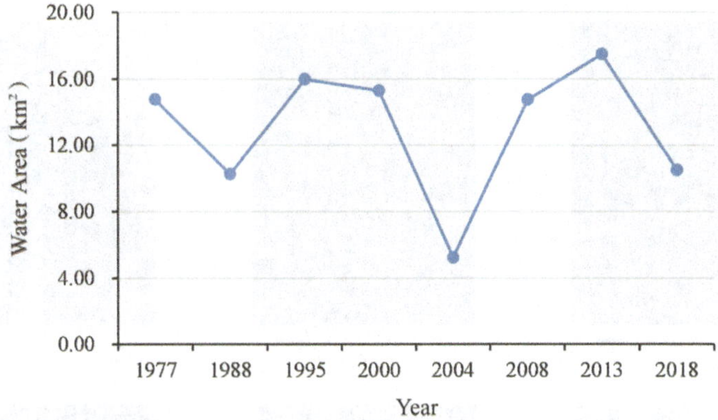

Fig. 128 Area dynamics of Kantale Tank

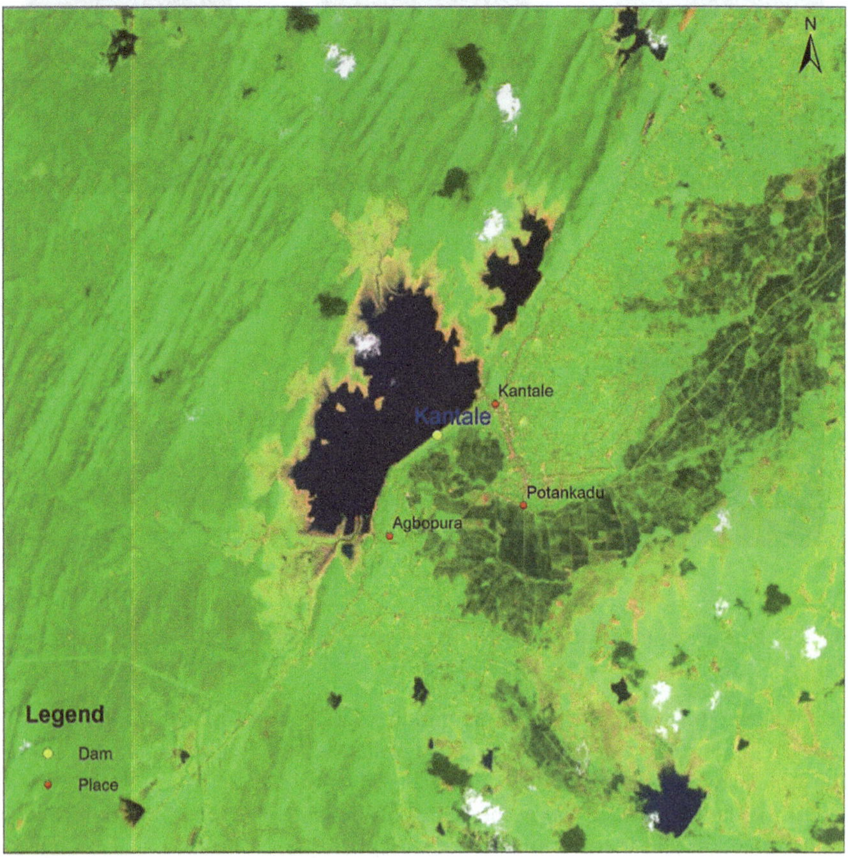

Fig. 129 True color remote sensing image of Kantale Tank

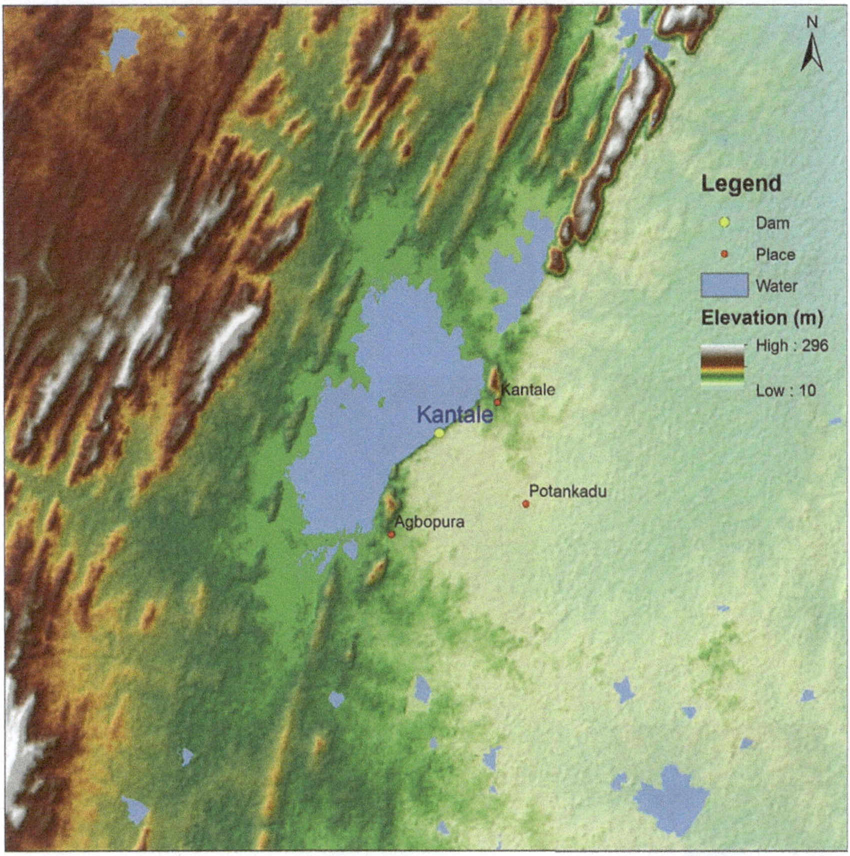

Fig. 130 Topography of Kantale Tank

30 years peak in at 27.23 km^2 in 2018 and the lowest (8.16 km^2) in 1992. From 2002 to 2018, the surface area of Uda Walawe Reservoir is on the rise (Figs. 132, 133, 134 and 135).

Parakrama Samudra

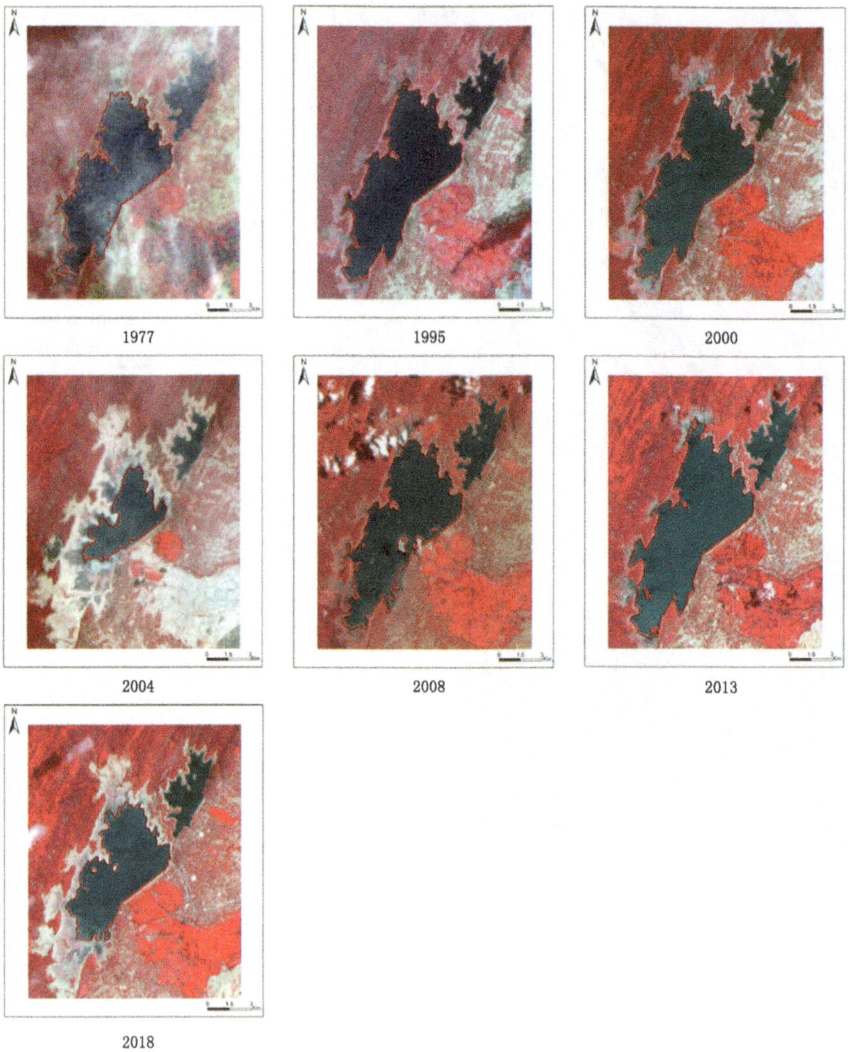

Fig. 131 Remote sensing image of Kantale Tank in different periods

Located in the north of Mahaweli River Basin and the west of Polonnaruwa, Parakrama Samudra is an ancient reservoir covering an area of 24 km^2 with 11 irrigation channels irrigating more than 16,000 ha of farmland. The bank of the reservoir is 13 km long and 12 m high with a catchment area of 75 km^2. The reservoir was built by King Parakramabahu and still plays an important role until today.

The water area information can be extracted from Landsat remote sensing images from 1977 to 2018. The results indicate that its surface area peaked at 21.14 km^2 in 1995 and reached the lowest at 9.33 km^2 in 2004. The surface area remained

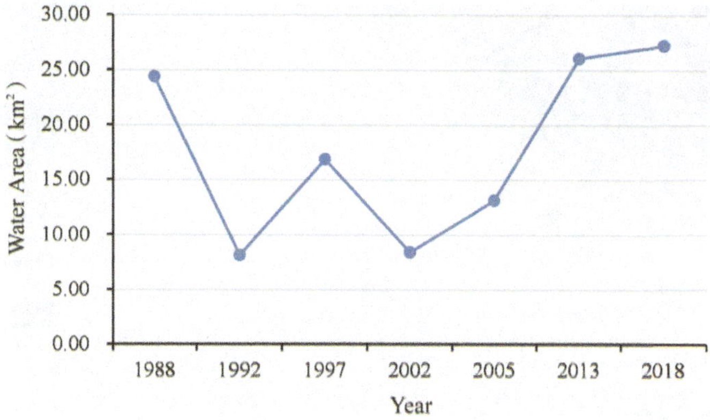

Fig. 132 Area dynamics of Uda Walawe Reservoir

Fig. 133 True color remote sensing image of Uda Walawe Reservoir

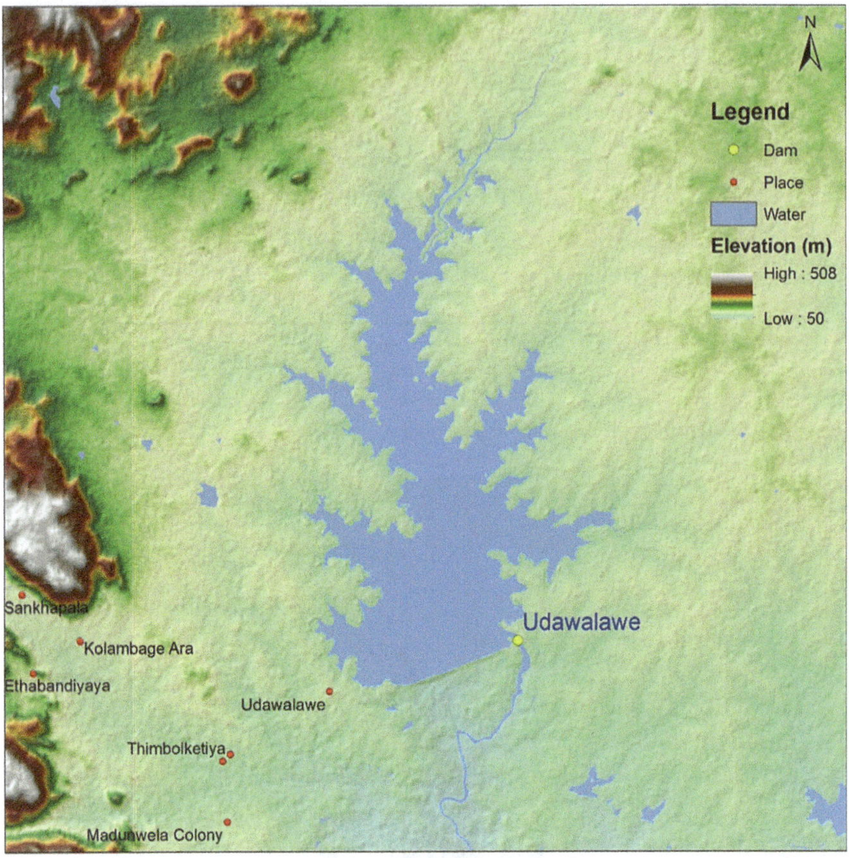

Fig. 134 Topography of Uda Walawe Reservoir

relatively more than 15 km^2 except for the year 2004 and thus little fluctuation in recent 40 years (Figs. 136, 137, 138 and 139).

Jayanthi Wewa Reservoir

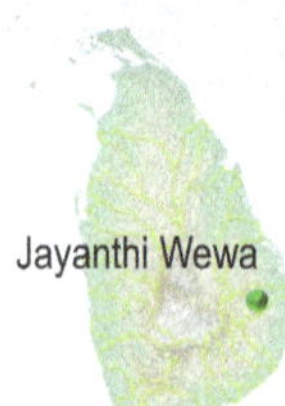

It is located the southeast of Senanayake Samudra along Gal Oya river in Eastern Province with a length of 23.44 km.

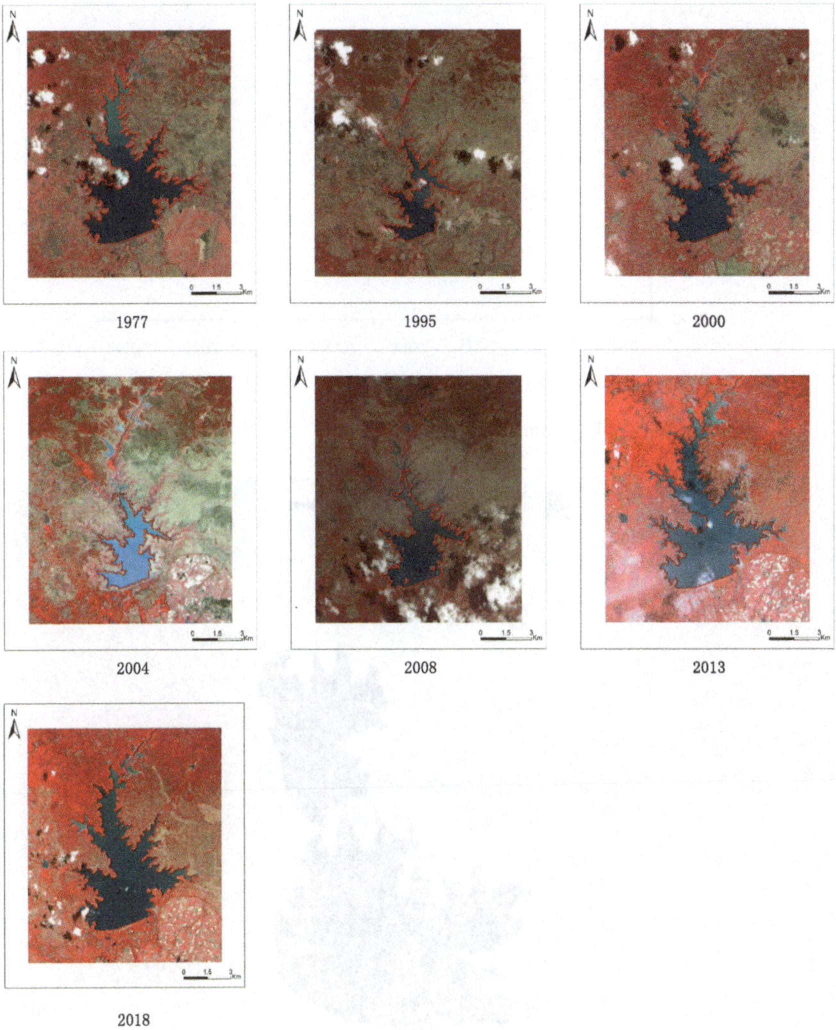

Fig. 135 Remote sensing image of Uda Walawe Reservoir in different periods

The water area information extracted from Landsat remote sensing images from 1977 to 2017 indicates that its surface area in the past 40 years fluctuated considerably peaking at 9.14 km^2 in 2012 from the lowest (1.97 km^2) in 1977. In general, the surface area of the reservoir significantly increased although with substantial variations (Figs. 140, 141, 142 and 143).

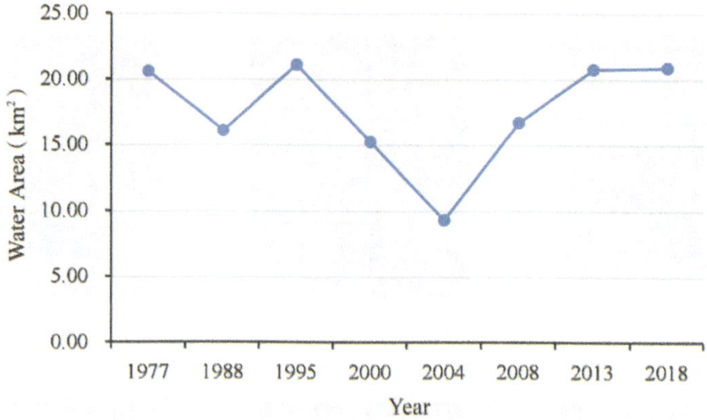

Fig. 136 Area dynamics of Parakrama Samudra

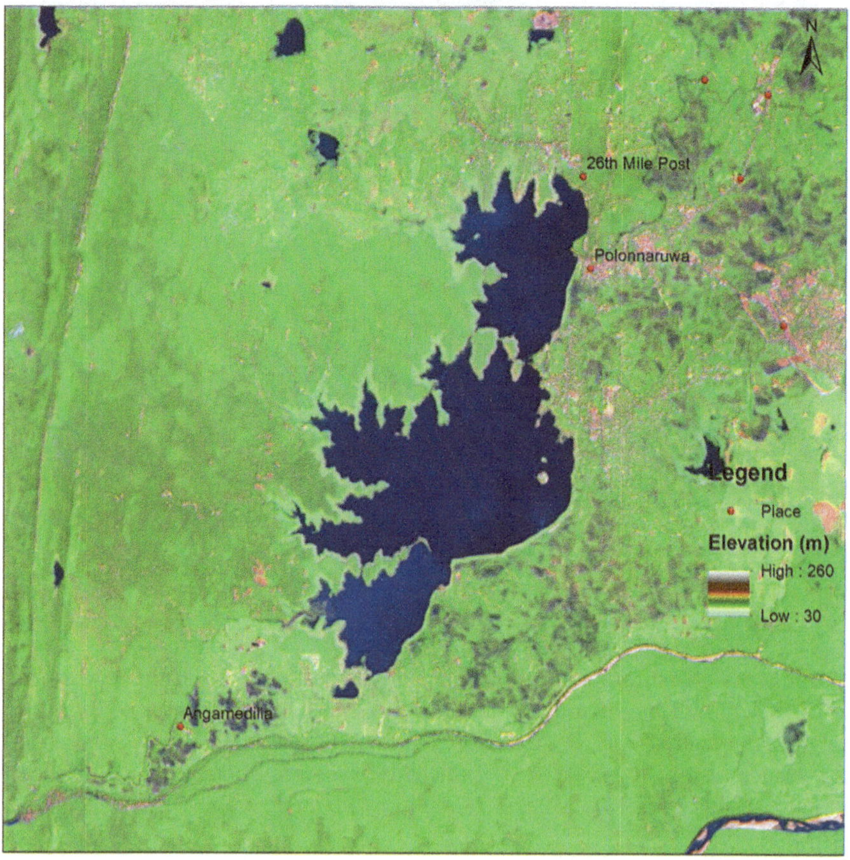

Fig. 137 Remote sensing image of Parakrama Samudra

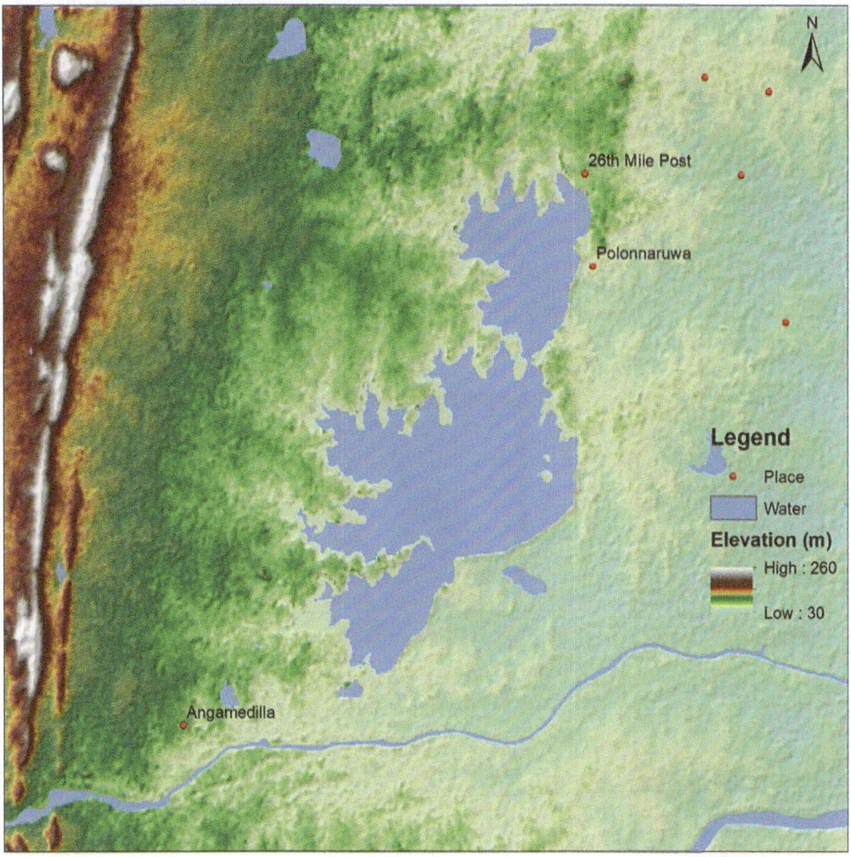

Fig. 138 Topography of Parakrama Samudra

Minneriya Wewa Reservoir

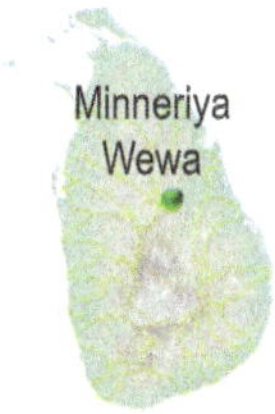

It is located between 8° 01′ N and 80° 52′ E, in the north of Mahaweli River Basin and about 16 km northwest of Polonnaruwa. Built in 3 BC, it has an area of 18.9 km^2.

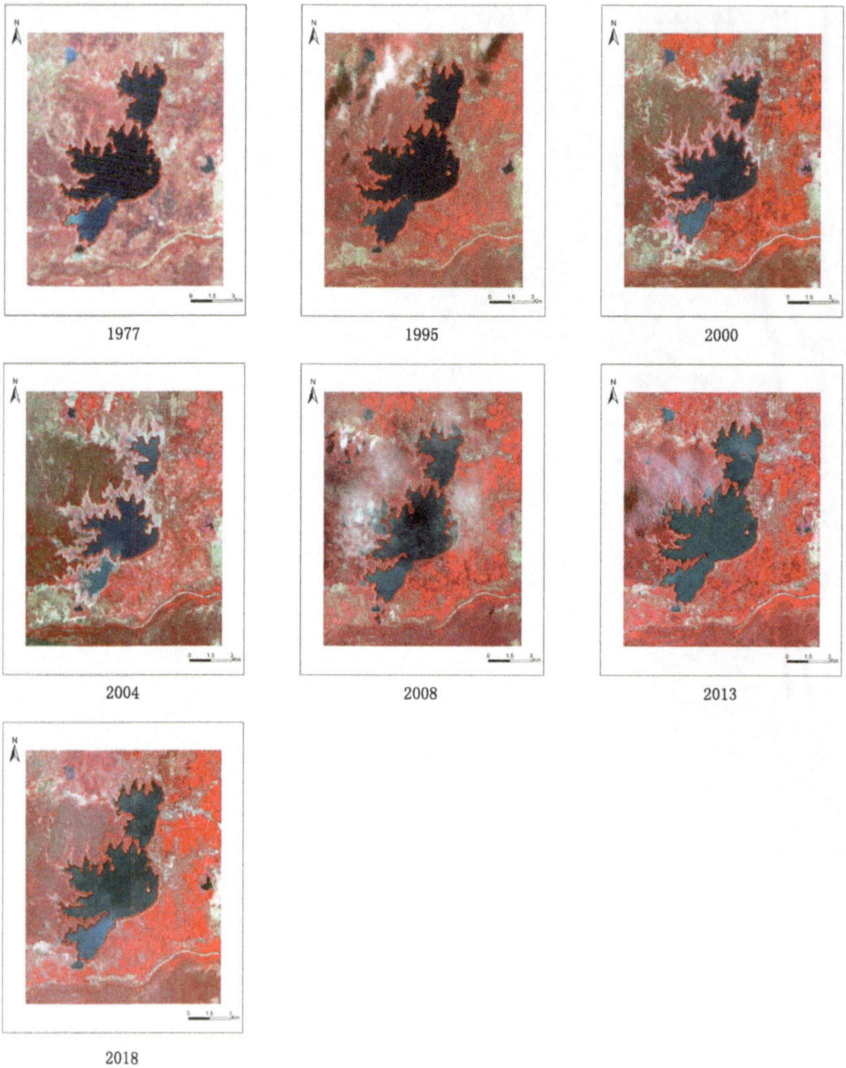

Fig. 139 Remote sensing images of Parakrama Samudra in different period

The water area information extracted from Landsat remote sensing images from 1977 to 2017 indicates that its surface area in the past 40 years fluctuated significantly peaking at 22.24 km^2 in 1977 and dipping to 6.97 km^2 in 2004. In general, the area of Minneriya Wewa Reservoir shows a downward trend although with substantial fluctuation (Figs. 144, 145, 146 and 147).

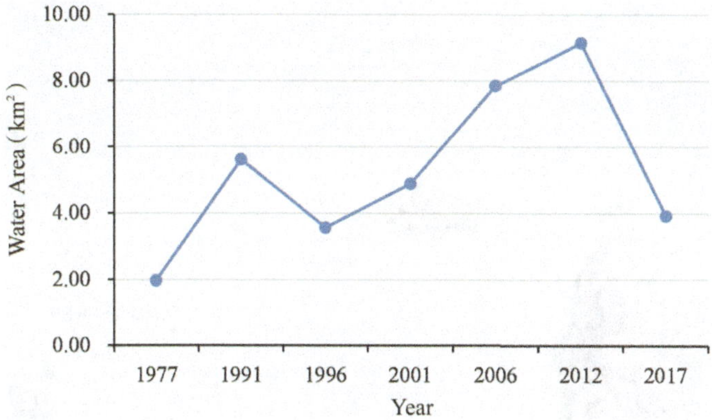

Fig. 140 Area dynamics of Jayanthi Wewa Reservoir

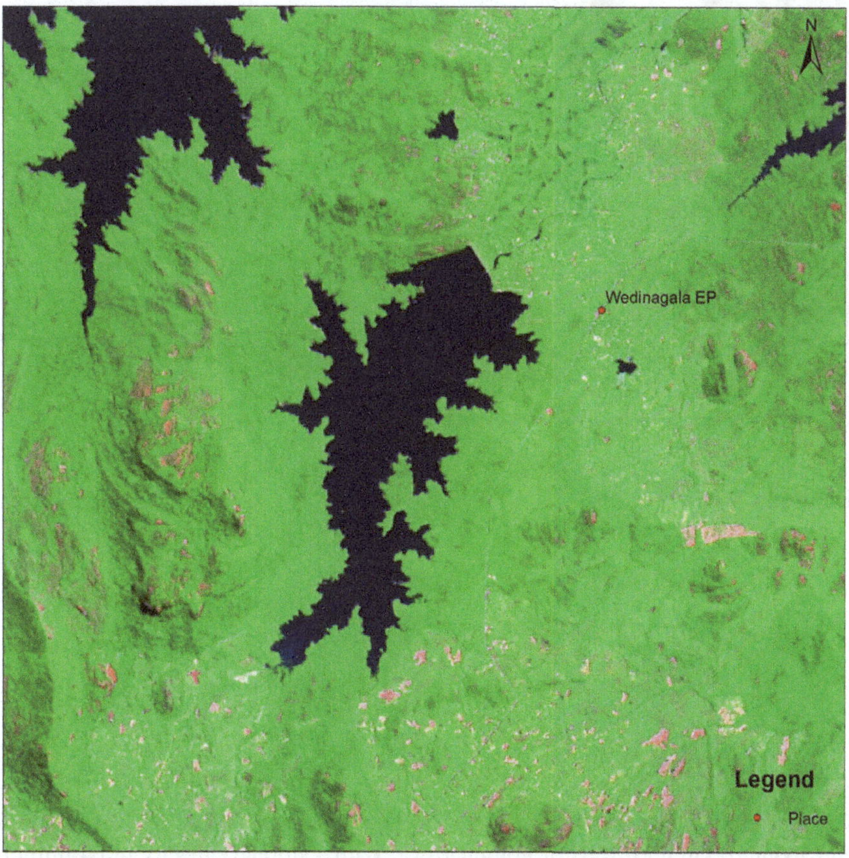

Fig. 141 True color remote sensing image of Jayanthi Wewa Reservoir

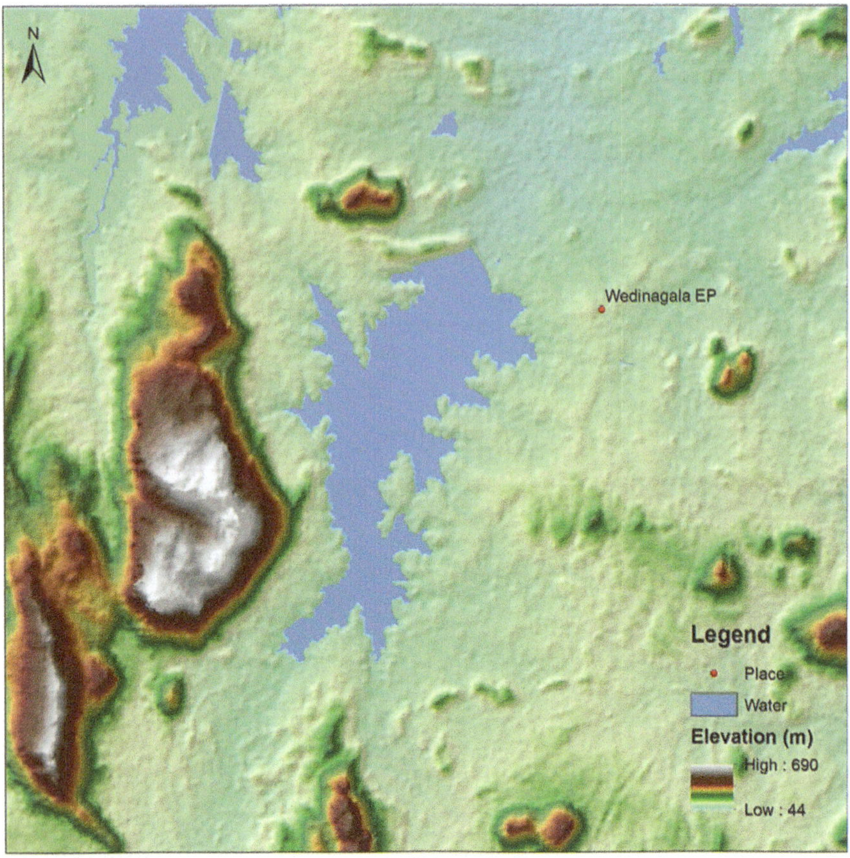

Fig. 142 Topography of Jayanthi Wewa Reservoir

Kaudulla Wewa

It is located between 8° 09′ N and 80° 55′ E, in the north of Mahaweli River Basin and about 20 km northwest of Polonnaruwa. With a storage capacity of 64 million cubic meters, Kaudulla Wewa Reservoir is 15 m high, 120 m wide and 9.2 km long.

The water area information obtained from Landsat remote sensing images from 1977 to 2018 indicates that its surface area peaked at 18.58 km^2 in 2013 from

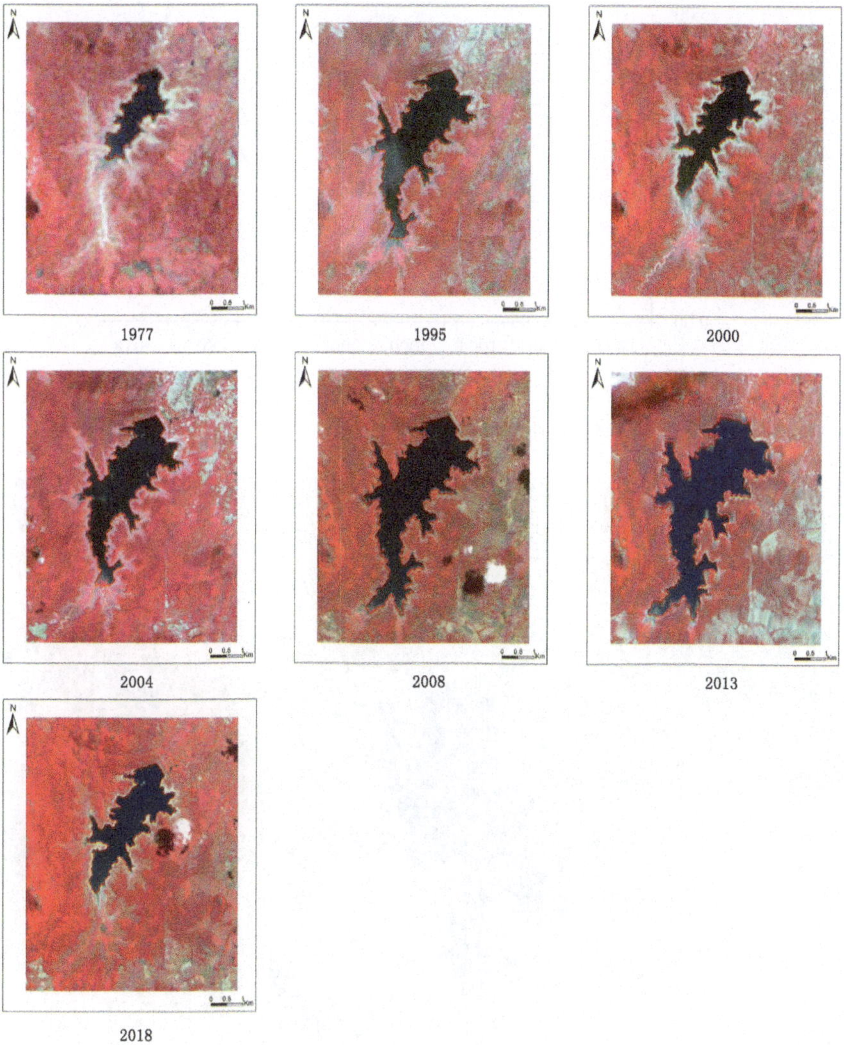

Fig. 143 Remote sensing image of Jayanthi Wewa Reservoir in different periods

the lowest dip of 6.09 km^2 in 2004. The area remained more than 12 km^2 except for the year 2004. Generally, the surface area undergoes small fluctuation despite occasionally sharp recess (Figs. 148, 149, 150 and 151).

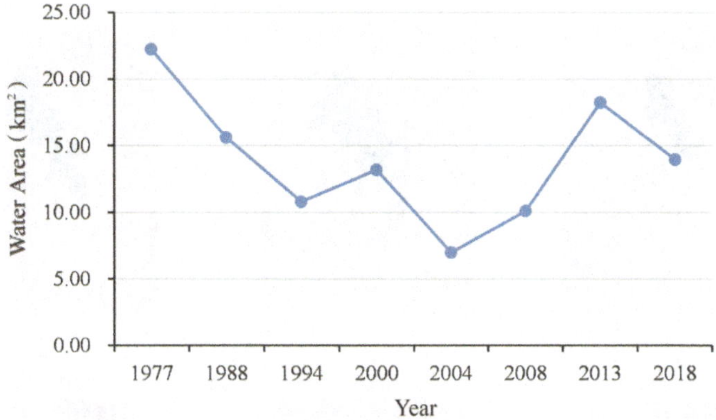

Fig. 144 Area dynamics of Minneriya Wewa Reservoir

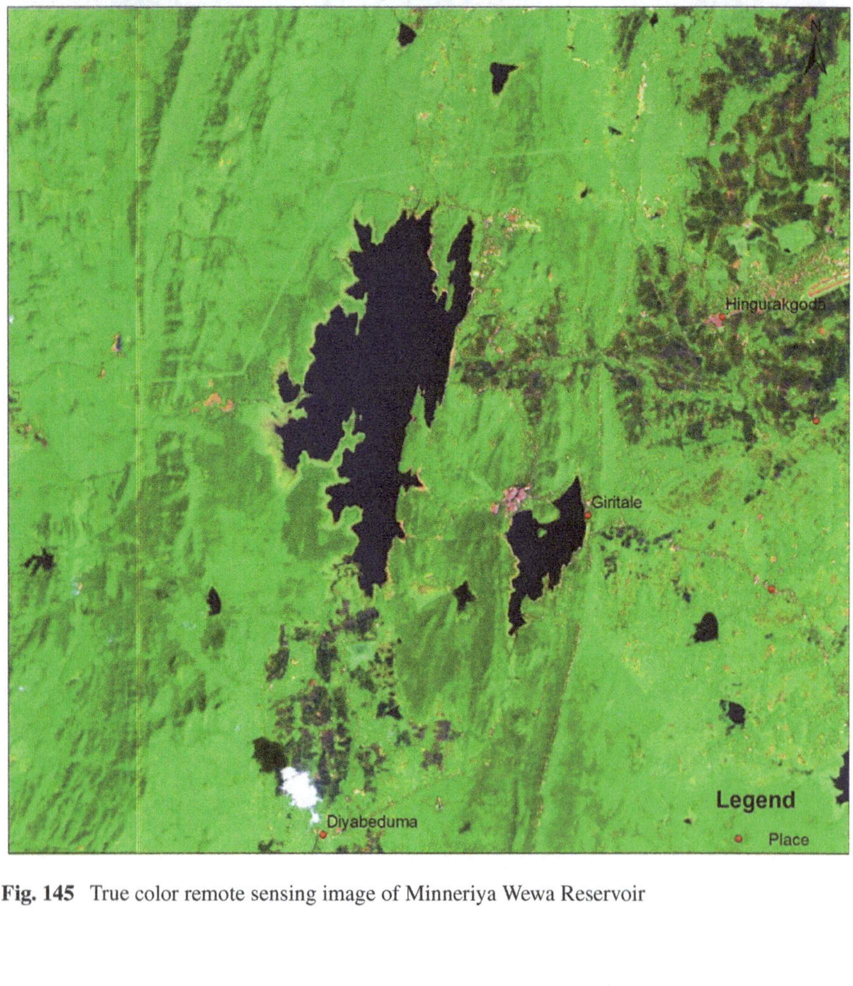

Fig. 145 True color remote sensing image of Minneriya Wewa Reservoir

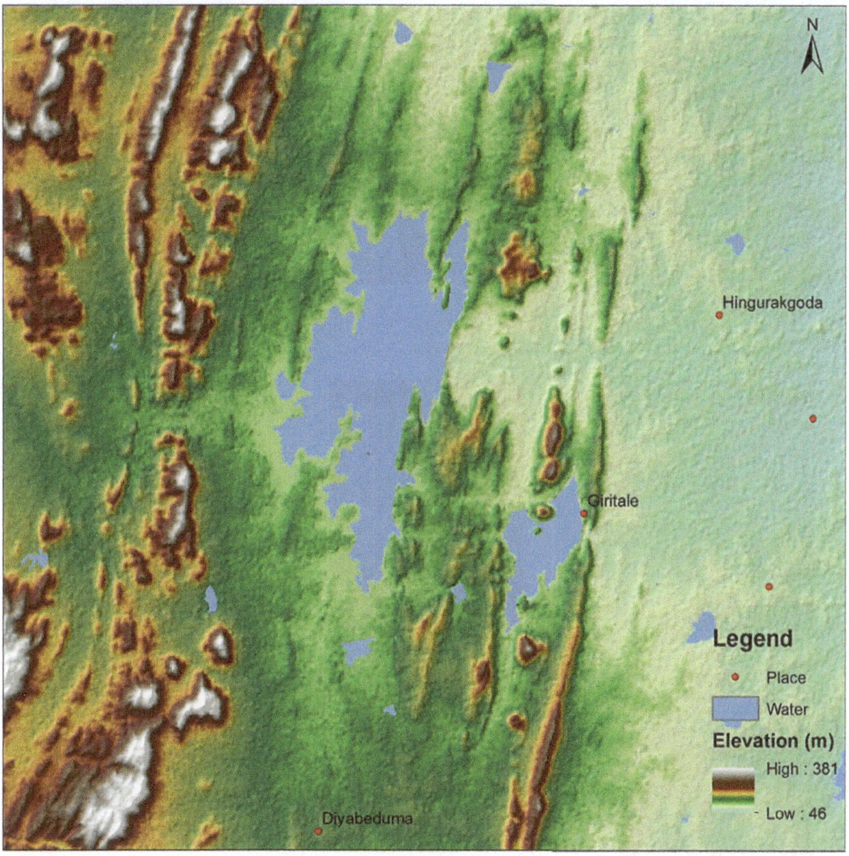

Fig. 146 Topography of Minneriya Wewa Reservoir

Irakkamam Wewa

At an altitude of 78 m, Irakkamam wewa is located in the lower reaches of the Gal Oya, the south-central part of Eastern Province between 8° 09′ N and 80° 55′ E.

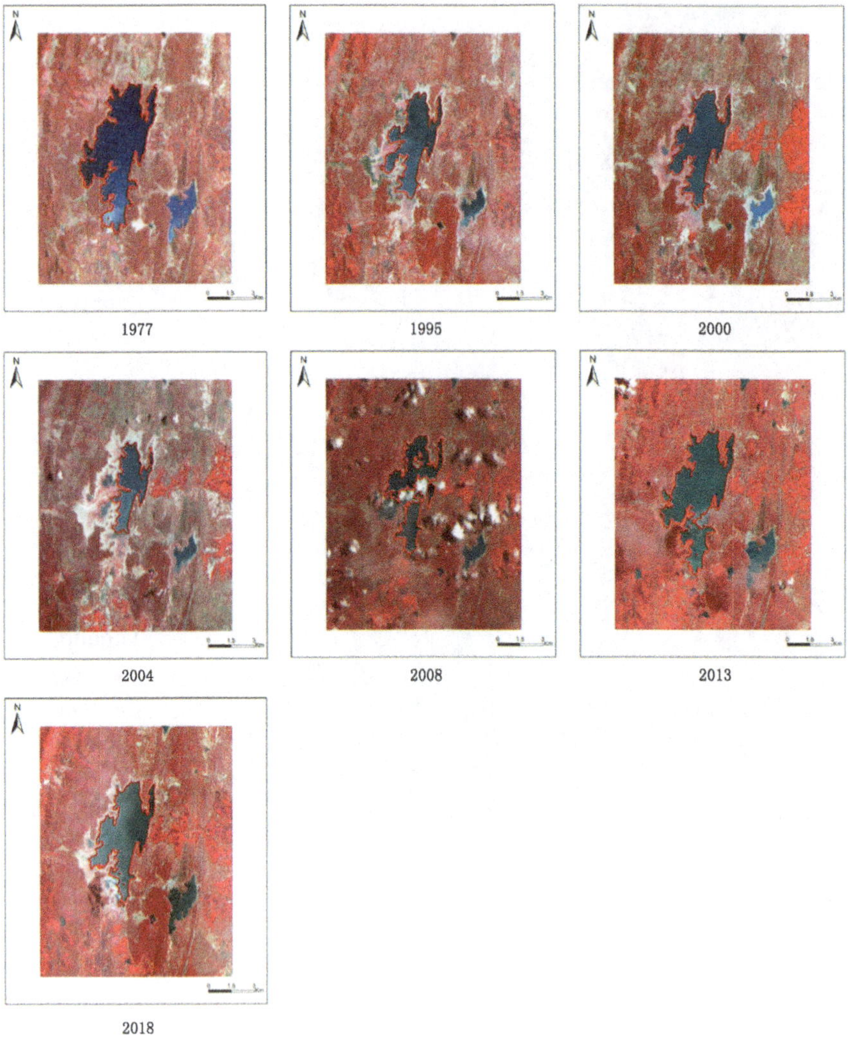

Fig. 147 Remote sensing image of Minneriya Wewa Reservoir in different periods

The water area information from Landsat remote sensing images from 1991 to 2017 indicates that its surface area fluctuated drastically peaking at 6.45 km^2 in 2012 from the lowest dip at 2.82 km^2 in 2006. The area remained relatively over 4 km^2 throughout except for the year 2006 (Figs. 152, 153, 154 and 155).

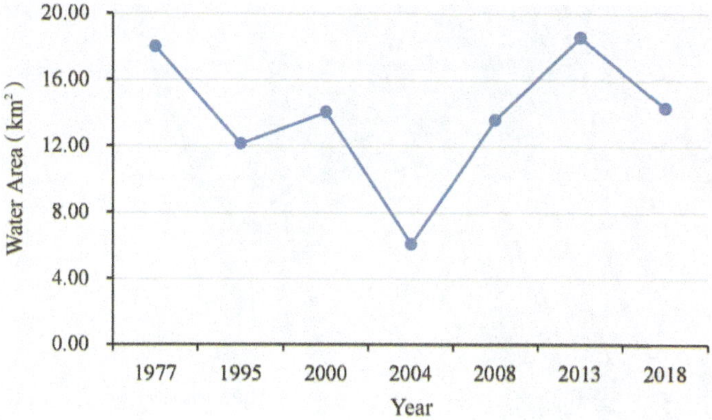

Fig. 148 Area dynamics of Kaudulla Wewa

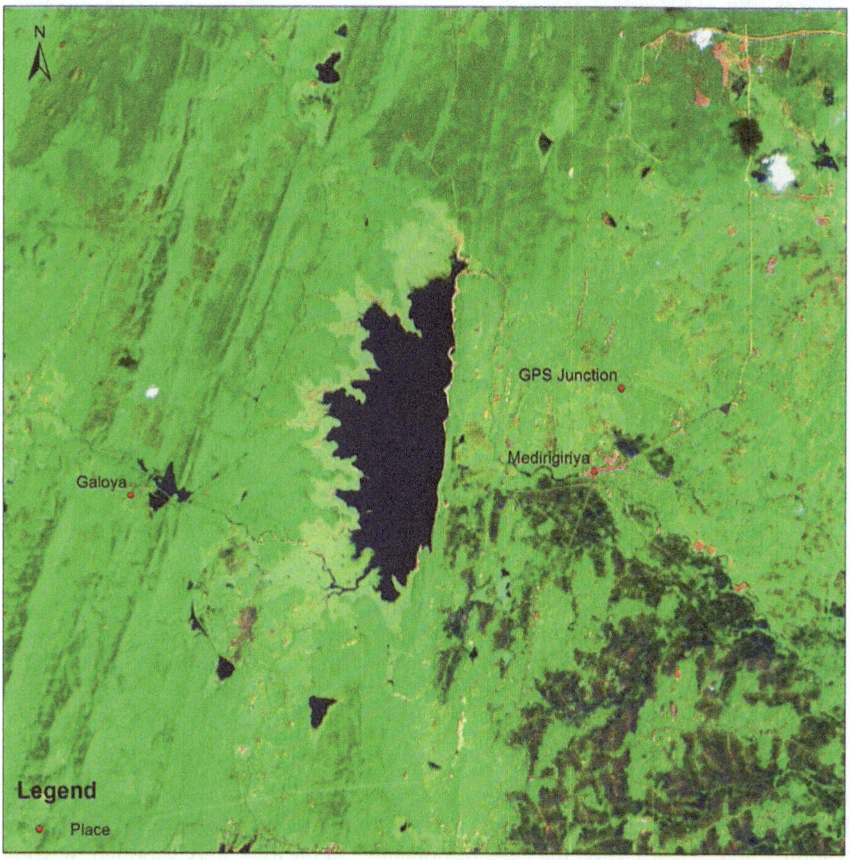

Fig. 149 True color remote sensing image of Kaudulla Wewa

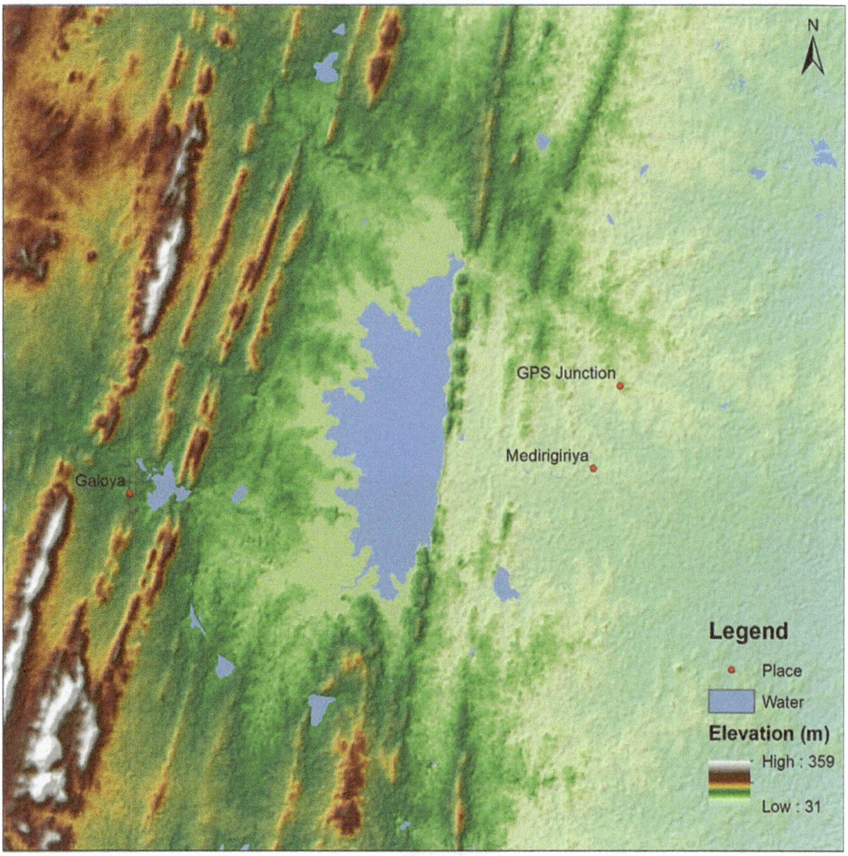

Fig. 150 Topography of Kaudulla Wewa

Nanthi Kadal Lagoon

It is located at 9° 17′ N and 80° 46′ E in Mullaitivu District, northeast Sri Lanka. Nanthi Kadal is said to be the "sea of conches", with its waters supplied by some small rivers such as Per Aru.

The water area information obtained from Landsat remote sensing images from 1988 to 2018 indicates that its surface area remained relatively unchanged over

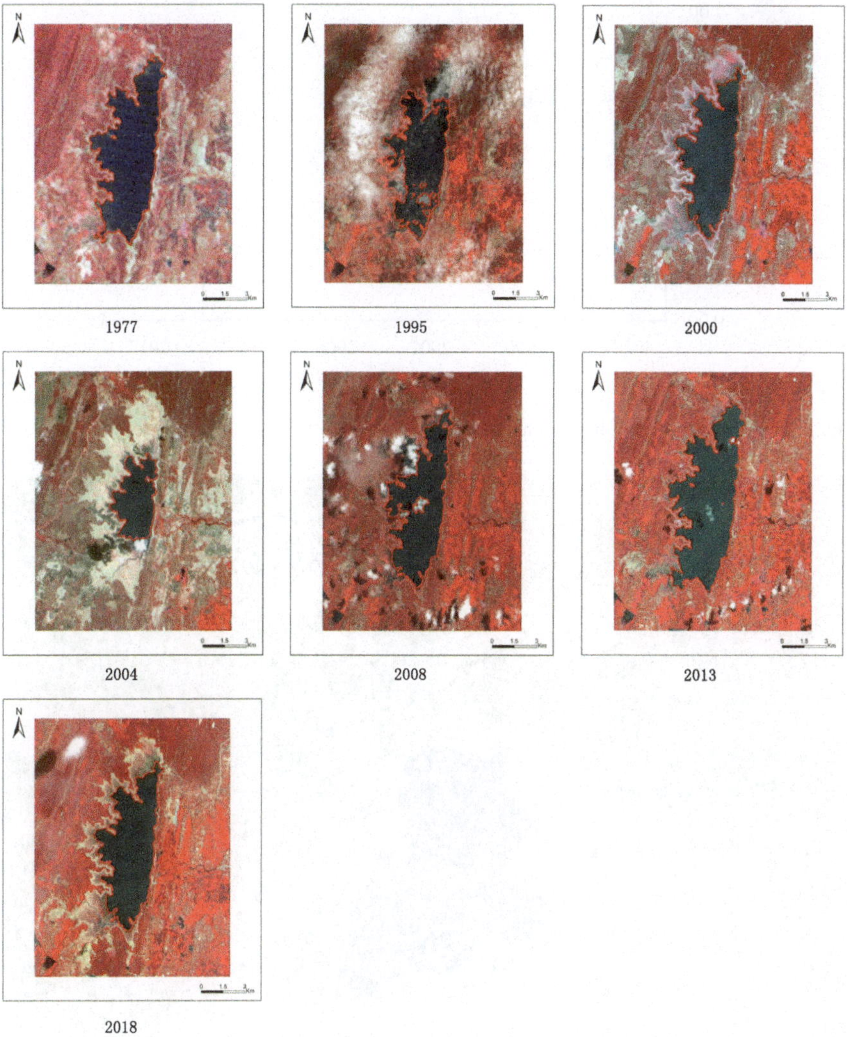

Fig. 151 Remote sensing image of Kaudulla Wewa in different periods

28 km^2 except for the year 2004 peaking at 30.59 km^2 in 2013 from the lowest dip of 25.58 km^2 in 2004. Generally, the reservoir surface area has experienced little change during this period, which is related to the irregular replenishment of sea water in monsoon season (Figs. 156, 157, 158 and 159).

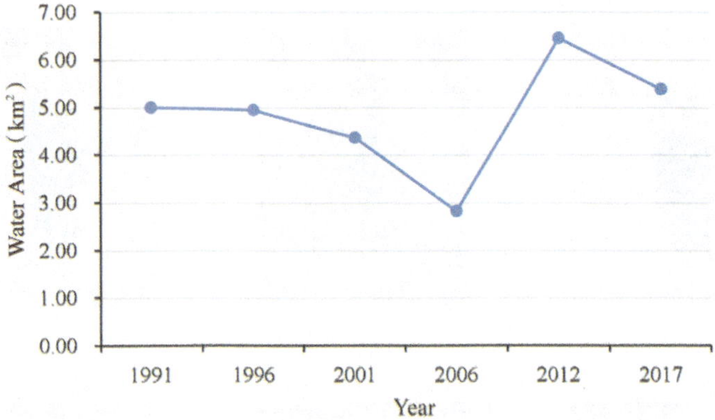

Fig. 152 Area dynamics of Irakkamam Wewa

Fig. 153 True color remote sensing image of Irakkamam Wewa

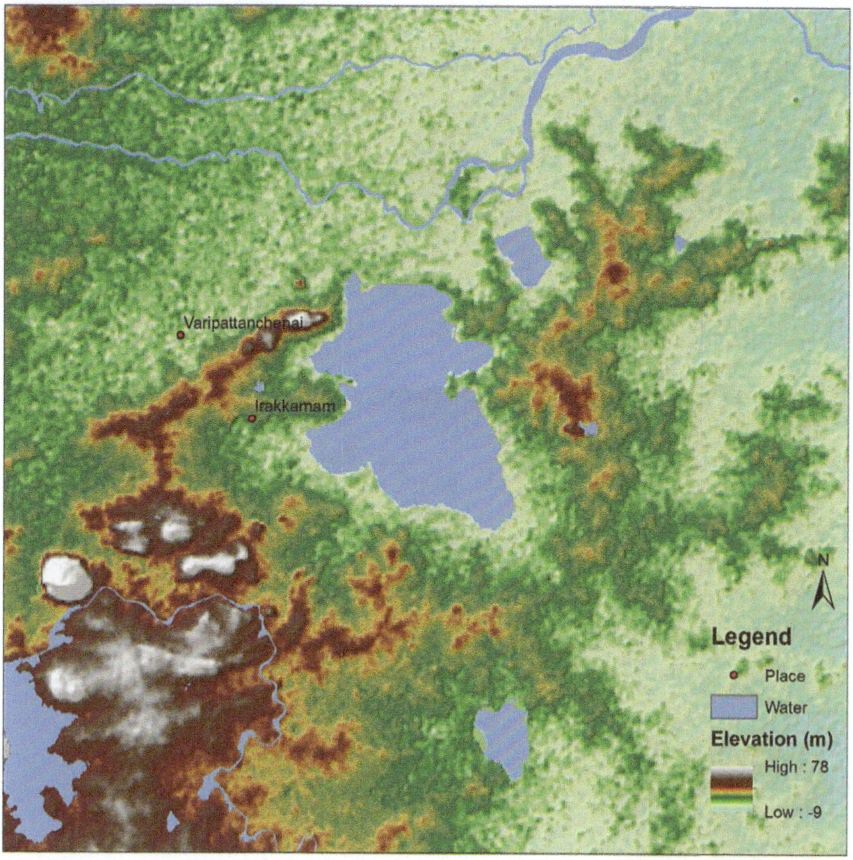

Fig. 154 Topography of Irakkamam Wewa

5 Marine Ecological Environment

To monitor and protect marine ecological environment in Sri Lanka, the distribution characteristics and dynamics of marine ecological environment under the dual pressures of natural changes and human activities will be analyzed based on the long-term stable monitoring data of marine remote sensing. The remote sensing monitoring data can show the basic status of marine phytoplankton for the protection of marine biodiversity and marine ecologically vulnerable areas as well as the near sea industry and fishery. It can also demonstrate the change of coastal water quality and offer support for the research of coastline migration, coastal engineering and its impact on coastal ecological environment. In addition, the long-term remote sensing data can be used

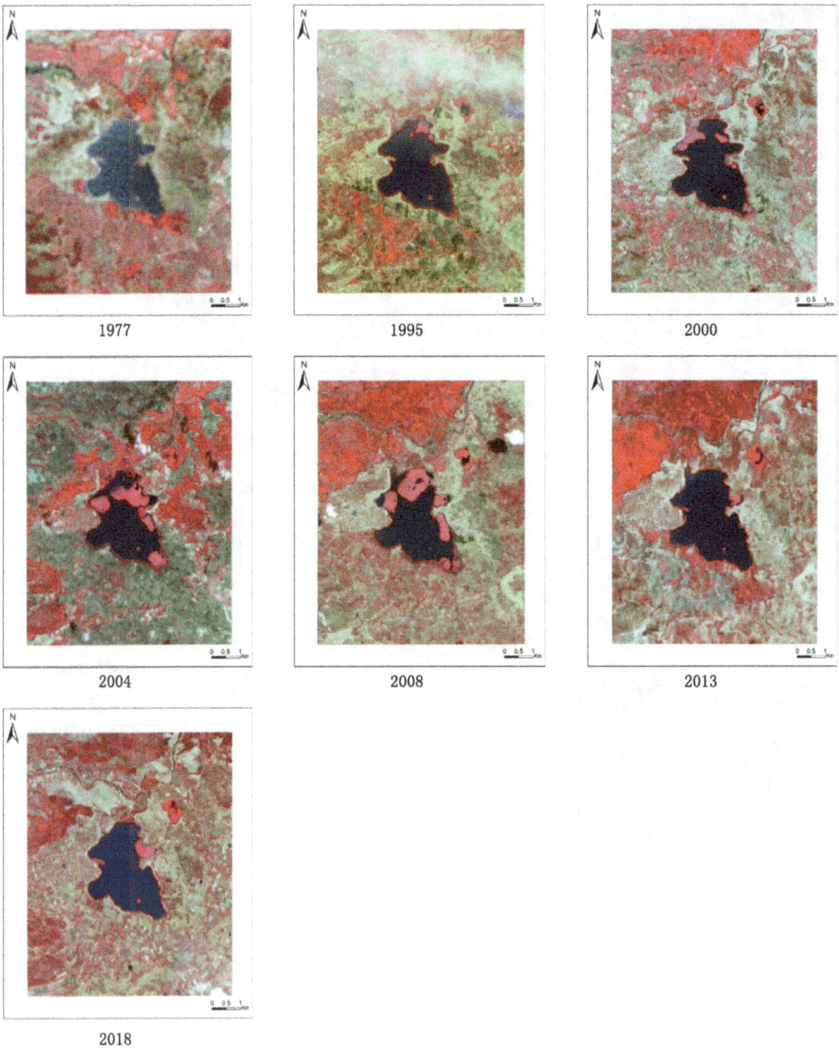

Fig. 155 Remote sensing image of Irakkamam Wewa in different periods

to monitor the interannual change of marine ecosystem, study the impact of environmental fluctuation on marine ecosystem under climate change and evaluate the impact of regional climate change disturbance on marine ecosystem such as El Nino phenomenon.

The main parameters concerned include water temperature (Sea Surface Temperature), sea surface light intensity (Photosynthesis Active Radiation), water cleanliness (Seawater Transparency), phytoplankton biomass and phytoplankton primary productivity as the basis of marine food web, etc.

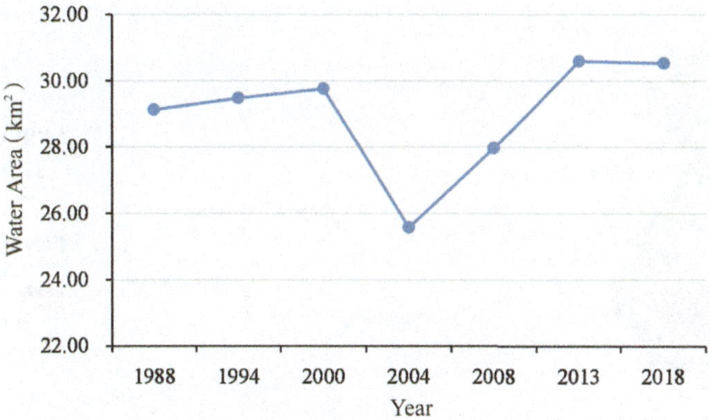

Fig. 156 Area dynamics of Nanthi Kadal Lagoon

Fig. 157 True color remote sensing image of Nanthi Kadal Lagoon

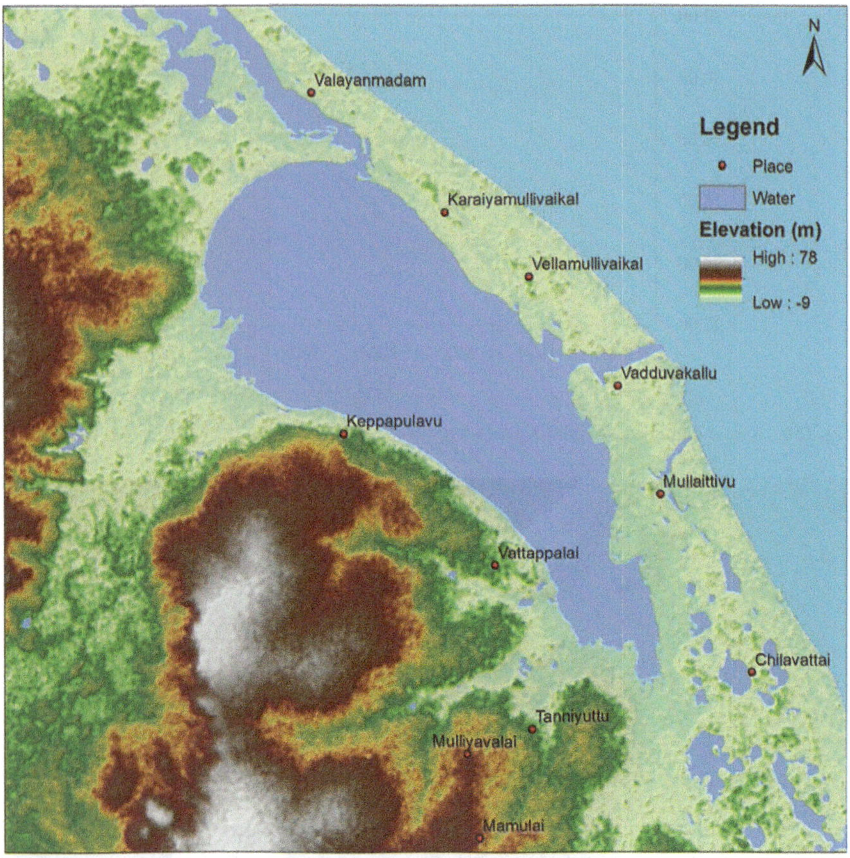

Fig. 158 Topography of Nanthi Kadal Lagoon

- Sea temperature (Sea Surface Temperature, SST)

It varies with latitude, depth and season. Sea temperature is an extremely important ecological restrictive factor for marine organisms. Through natural selection, the adaptation of each marine organism to water temperature has a specific range. That means each has the lowest, highest and optimum temperature that can be tolerated, and the lowest and highest temperature required in the growth, development and reproduction stages. Therefore, it is the main environmental factor that determines the habitat, species richness and the variation of marine organisms.

- Light (Photosynthesis Active Radiation, PAR)

Sunlight is the main energy source of the earth's ecosystem. It is only through the photosynthesis that the solar energy can be continuously assimilated into the ecosystem and ultimately used by other organisms. PAR can be used to express the

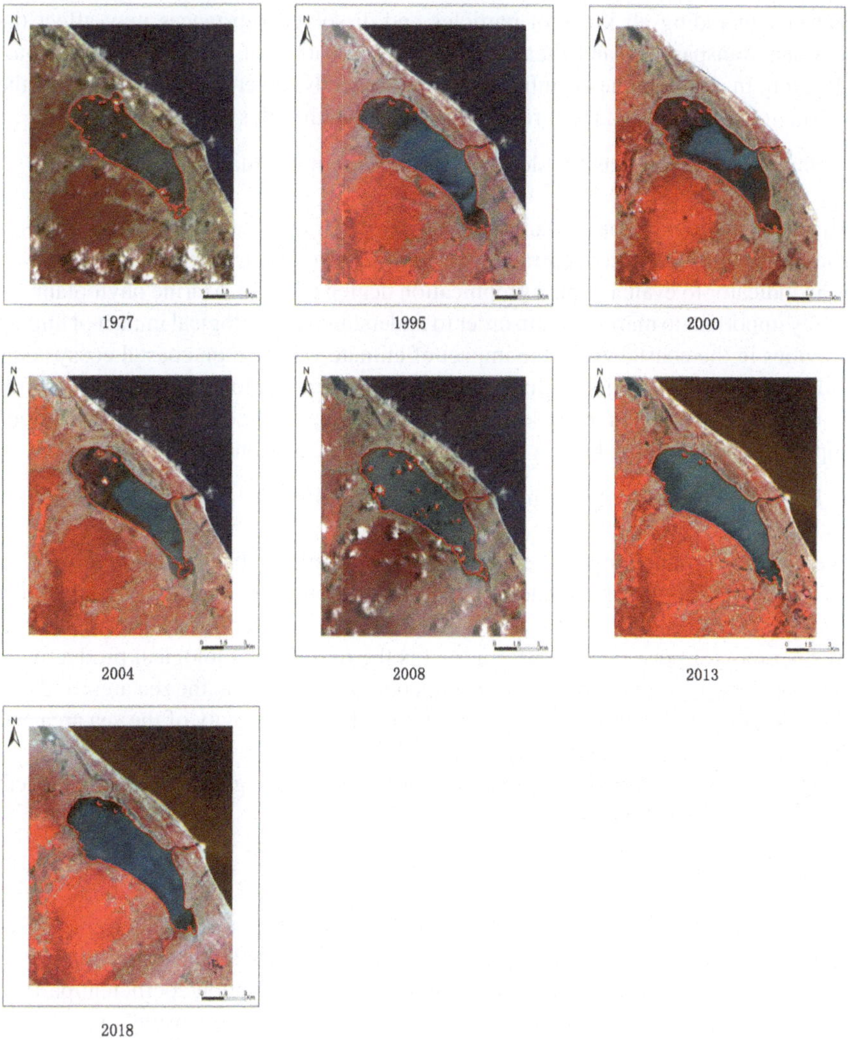

Fig. 159 Remote sensing image of Nanthi Kadal Lagoon in different periods

solar energy received and utilized by phytoplankton. It is the part of spectral energy absorbed by phytoplankton in the process of photosynthesis. The general wavelength is between 380 and 710 nm.

- Water cleanliness (seawater transparency, SDD)

It is the same as seawater transparency. Usually, the transparency plate is used to measure the cleanliness, which can sink into the water vertically where the sunlight cannot directly shine until the invisible depth. The content of substances in the seawater including all kinds of particles and dissolved substances may affect the seawater transparency and then affect the illumination of the underwater phytoplankton. In addition, the cleanliness will also directly affect the water quality of the coastal tourism area and the direct sensory and health status of human being.

- Phytoplankton biomass (Chlorophyll Concentration, Chla)

Almost all phytoplankton contain chlorophyll as their main carrier of photosynthesis. Therefore, chlorophyll concentration is usually used to characterize biomass and used as an indicator to evaluate the eutrophication degree of water. Marine phytoplankton is very important to marine life. In order to understand the ecological impact of human activities in the coastal zone, the impact of climate variation on coastal ecosystems and the causes of variation of fishery and other marine biological resources, phytoplankton dynamics must be assessed. Harmful algal may exert increasingly serious impacts on aquaculture, human health and coastal ecosystems in many areas.

- Biological productivity (Primary Productivity, NPP)

The Primary Productivity is the ability of phytoplankton, benthos and autotrophic bacteria to produce organic matter and fix energy through photosynthesis or chemical synthesis. It is generally expressed by the fixed organic carbon or energy per unit area per day (or per year), gC m^{-2} a^{-1}. It is the most basic biological productivity, the basis for the production of organic or economic products in the sea area, and one of the important indicators for the estimation of the productivity of the sea area and the potential of fishery resources.

The data and methods for remote sensing monitoring of marine ecological environment are as follows:

(1) **Data Source**

The original satellite data used are Aqua/MODIS remote sensing products, including the monthly average chlorophyll concentration, seawater temperature, photosynthesis available radiation, remote sensing reflectance/absorption coefficient/particle backscatter coefficient of 443 nm water body, and the monthly average marine net primary production of 9 km resolution published by the University of Oregon Strength and chlorophyll concentration (for modifying the primary productivity model). The seawater temperature is the nighttime temperature retrieved from the thermal infrared band. Moreover, the remote sensing reflectance, absorption coefficient and particle backscatter coefficient of 443 nm of the water body are added to further invert the transparency of sea water and correct the original NPP remote sensing products. On the basis of monthly remote sensing products, 15 years (2002–2017) fitting regression is carried out to get the change rate of various elements in different regions (Table 11).

Table 11 Original satellite data products

Remote sensing data name	Sensor	Satellite	Period	Time resolution	Spatial scope	Spatial resolution	Source
Chlorophyll concentration	MODIS	Aqua	2002–2017	Monthly average	Globe	9,000 m	NASA[a]
Seawater temperature	MODIS	Aqua	2002–2017	Monthly average	Globe	9,000 m	NASA
Photosynthesis active radiation	MODIS	Aqua	2002–2017	Monthly average	Globe	9,000 km	NASA
Water remote sensing reflectance (443 nm)	MODIS	Aqua	2002–2017	Monthly average	Globe	9,000 km	NASA
Water absorption coefficient (443 nm)	MODIS	Aqua	2002–2017	Monthly average	Globe	9,000 km	NASA
Particle backscatter coefficient (443 nm)	MODIS	Aqua	2002–2017	Monthly average	Globe	9,000 km	NASA
Primary productivity	MODIS	Aqua	2002–2017	Monthly average	Globe	9,000 km	University of Oregon[b]
Chlorophyll concentration (for correcting primary productivity)	MODIS	Aqua	2002–2017	Monthly average	Globe	9,000 km	University of Oregon

[a]Website: http://oceandata.sci.gsfc.nasa.gov/MODISA/Mapped/Monthly/9km
[b]Website: http://orca.science.oregonstate.edu/2160.by.4320.monthly.hdf.vgpm.m.chl.m.sst.php

(2) **Methods**

- Seawater temperature, chlorophyll concentration and production of photosynthesis active radiation

Based on the original remote sensing data of global sea surface temperature, chlorophyll concentration and photosynthesis active radiation, the selected regions are resampled to generate the remote sensing data of sea surface temperature, chlorophyll concentration and photosynthesis active radiation.

- Seawater transparency

Global seawater transparency can be retrieved through semi analytical remote sensing model:

$$SDD = \frac{1}{4(a + b_b)} \ln\left(\frac{\rho_d \alpha \beta}{C_e R}\right) \tag{5}$$

SDD is the seawater transparency. a and b_b are the total absorption coefficient and backscatter coefficient of water body respectively. From the calculation, α and β can be obtained as the refraction and reflection influence coefficients of water surface ($\alpha\beta \approx 0.15$). ρ_d is the surface reflectance (~0.82) on the transparency disk, C_e is the contrast threshold (~0.02) of human eyes and R is the albedo of water subsurface, which can be calculated with remote sensing reflectance of water surface.

$$R = \frac{QR_{rs}}{0.52+1.7R_{rs}} \tag{6}$$

Q is the ratio (~4.0) of the ascending and the ascending irradiation of the water subsurface and R_{rs} is the acquired 443 nm remote sensing reflectivity.

- Primary Productivity

The standard NPP retrieved from the VGPM algorithm is significantly overestimated in the turbid water near the shore. This is because the algorithm only uses chlorophyll concentration to estimate the depth of the true light layer without the impact of land-based suspended solids, resulting in the overestimation of the depth of the true light layer. Therefore, the depth of the true light layer needs to be corrected to generate the global NPP remote sensing. The correction is as follows:

In the standard VGPM algorithm, the estimation model of the true light layer depth (Z_{eu}) is:

$$Z_{eu} = 200 \times Chl_t^{-0.293} \tag{7}$$

Chl_t is water column chlorophyll, which can be estimated by remote sensing surface chlorophyll concentration:

$$\begin{cases} Chl_t = 38.0 \times Chl^{0.425}, 当 Chl \le 1.0\, \mu g/l \\ Chl_t = 40.2 \times Chl^{0.507}, 当 Chl > 1.0\, \mu g/l \end{cases} \tag{8}$$

The seawater transparency can be retrieved to estimate the true light layer depth:

$$Z_{eu}(SDD) = 2.63 \times SDD \tag{9}$$

Finally, we can get the corrected NPP:

$$NPP_{cor} = NPP_{VGPM} \times Z_{eu}(SDD) / Z_{eu} \tag{10}$$

NPP_{VGPM} is the original net primary productivity and NPP_{COR} is the corrected one.

(3) **Statistical methods**

- Remote sensing data with monthly, seasonal and annual average distribution of sea climate

According to the specific sea area (the whole sea area or key area), the monthly remote sensing data of 15 years (2002–2017) were used to calculate the multi-year monthly, seasonal and annual average value of each pixel in the sea area for final date collection. The division method of each season: the first quarter is from March to May, the second is from June to August, the next is from September to November and the final is from December to February.

- The monthly and yearly average value of sea climate

On the basis of the monthly and yearly average remote sensing data of the sea area climate, the monthly and annual average values of the sea area climate can be obtained through the arithmetic average of the effective pixel values.

- Yearly change

On the basis of monthly average remote sensing data of sea climate, the yearly change (maximum minus minimum) can be calculated for the remote sensing data of yearly change distribution by subtracting the minimum value from the maximum one of each image element in 15 months.

- Change velocity

The velocity can be obtained through the average value from the linear regression of 15-year monthly average data, which is also the ratio of slope to average value.

5.1 Distribution and Changes of Marine Ecological Environment Surrounding Sri Lanka

Multiyear average distribution characteristics of ecological environment factors in the sea area around Sri Lanka is as follows.

The recent 15-year average distribution of ecological environment obtained from remote sensing images in 2002–2017 shows that the annual average seawater temperature of the sea area around Sri Lanka is about 28 °C while the near sea temperature is slightly lower than that of the outer sea. Except for the relatively low annual average photosynthesis active radiation (PAR) around the island, the offshore figure is higher than that of the outer sea due to relatively lower cloud coverage. Influenced by the input of land-based materials, human activities and the sediment resuspension in the shallow water, the transparency of coastal and offshore water is relatively low, with a minimum transparency of less than 2 m. Nevertheless, the transparency of the open sea is quite high with an annual average of more than 30 m. The biomass of phytoplankton in the coastal area is relatively high with more than 2 μg/L of the average annual chlorophyll concentration, but the range of high phytoplankton biomass is narrow. The biomass of phytoplankton in the outer sea is extremely low with less than 0.2 μg/L of the average annual chlorophyll concentration. Similarly, the NPP value is generally low except for that of the nearshore, mostly close to 500 mgC/(m^2 day) (Fig. 160).

In terms of annual distribution characteristics of ecological environment factors in the sea area around Sri Lanka, the water temperature remains relatively constant below 3 °C, especially in the southern sea area with the temperature of less than 2 °C. The annual variation of photosynthesis active radiation is lower than 20 Einstein/(m^2day) and the Northeast sea area has a relatively large annual variation. The maximum annual variation of seawater transparency is reflected in the offshore area reaching more than 25 m, while that of coastal and open sea is as small as less than 10 m, especially in the coastal area with depth less than 5 m. The annual variation range of phytoplankton biomass and primary productivity is the highest in the southern coast. The change of chlorophyll concentration reaches 10 μg/L and the annual change of primary productivity can reach 3000 mgC/(m^2day) (Fig. 161).

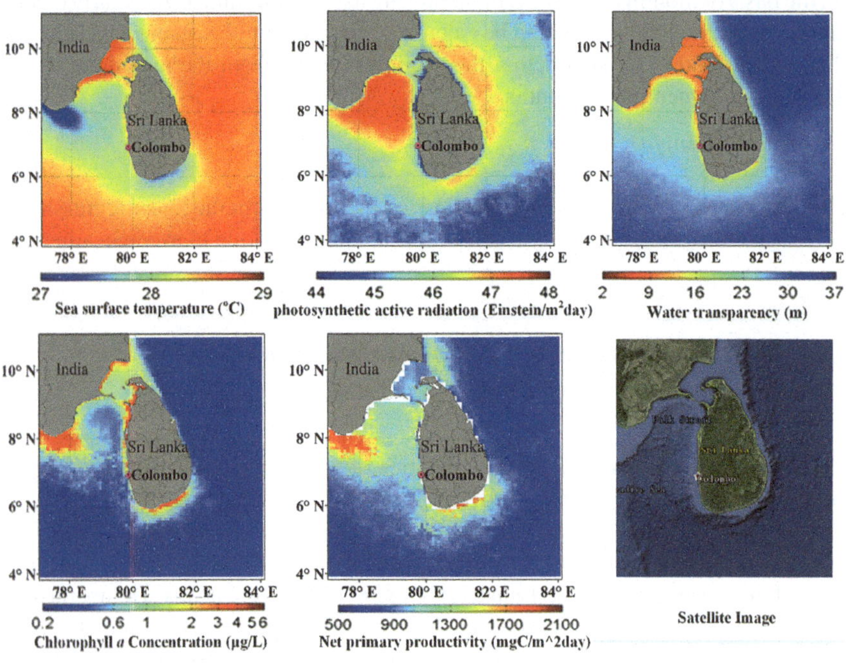

Fig. 160 Annual average (2002–2017) spatial distribution of ecological environment factors in the sea area around Sri Lanka, such as sea surface temperature, photosynthesis active radiation, seawater transparency, chlorophyll concentration and primary productivity)

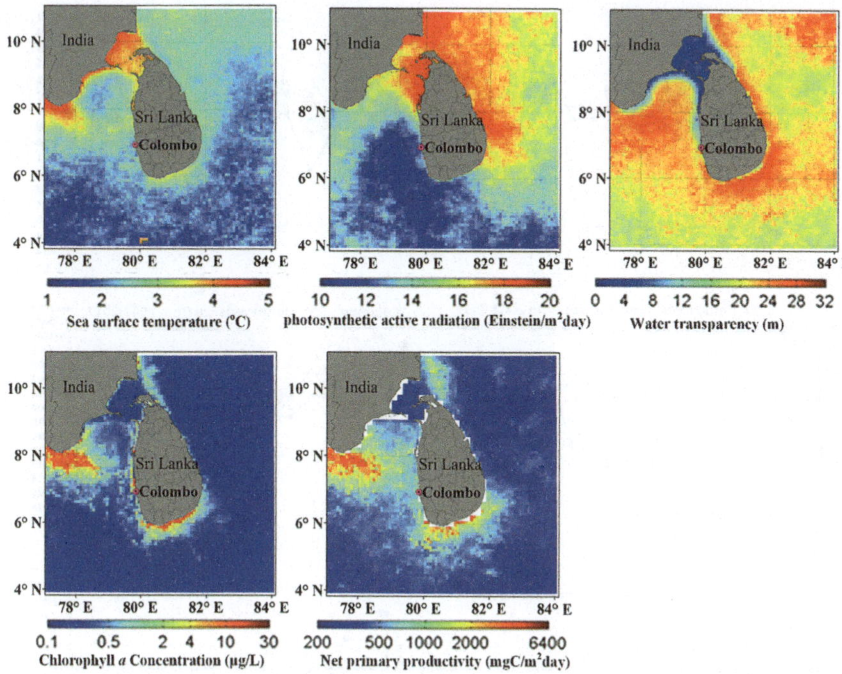

Fig. 161 Annual variation of environmental factors in the sea area around Sri Lanka, such as sea surface temperature, photosynthesis active radiation, sea water transparency, chlorophyll concentration and primary productivity

5.2 Interannual and Intermonthly Changes of Marine Ecological Environment Surrounding Colombo

From 2002 to 2017, the highest seawater temperature (SST) in the sea area around Colombo appeared in April and the highest photosynthesis active radiation (PAR) in March. The transparency in the first half of the year was significantly better than that in the second half of the year, while the productivity in the second half of the year was significantly higher than that in the first half of the year. In recent years, the concentration of phytoplankton chlorophyll has been higher in August, hence the possibility of red tide outbreak in this month. The SST, PAR and chlorophyll concentration (Chla) all showed an upward trend, with an annual increase rate of 0.11%, 0.08% and 1.89% respectively. The transparency and primary productivity of seawater decreased at the rate of 0.08% and −0.80% respectively (Figs. 162 and 163).

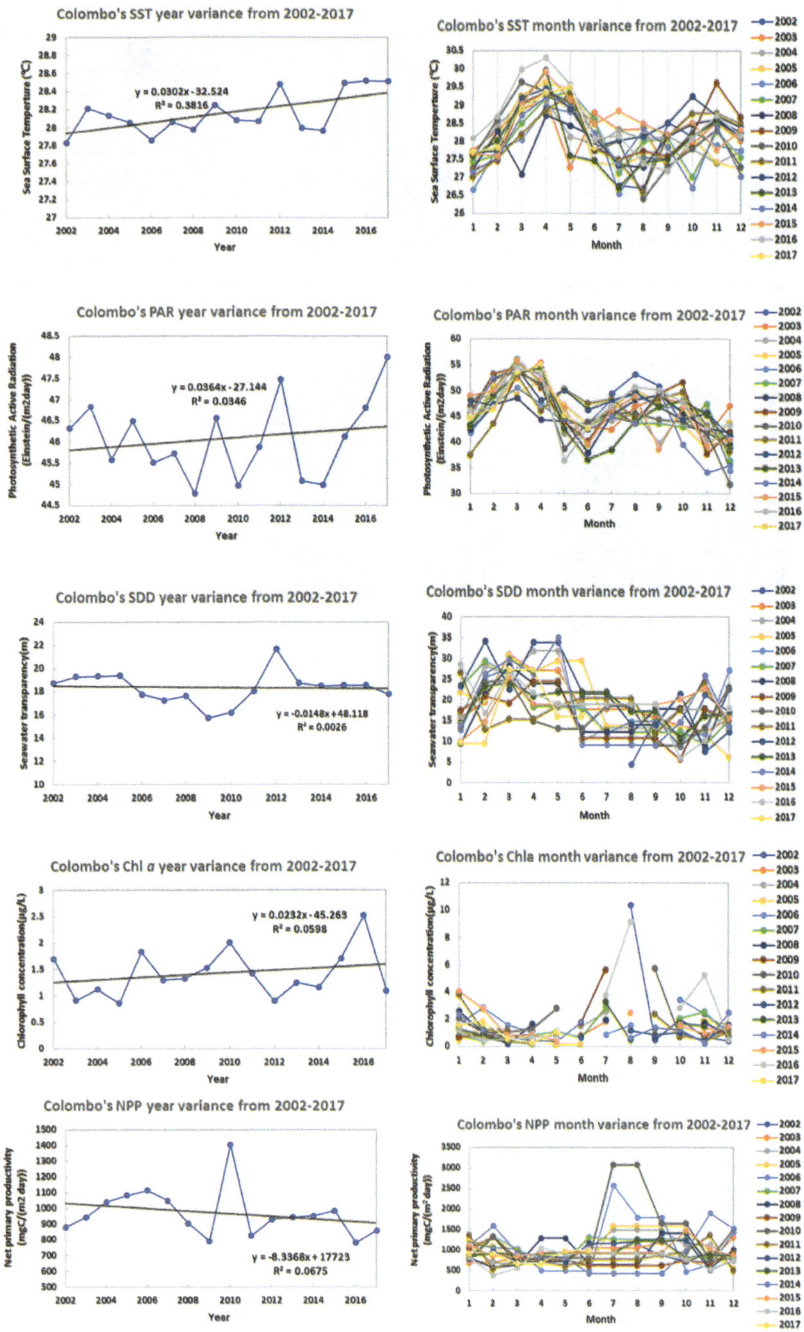

Fig. 162 Interannual and intermonthly changes of ecological environment factors in the sea area around Colombo

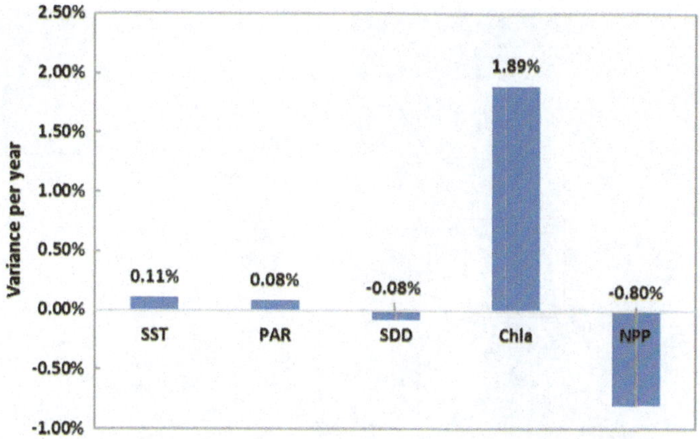

Fig. 163 Change rate of ecological environment factors in the sea area around Colombo port (% per year)

5.3 Spatio-Temporal Changes of Ocean Bio-Optical Properties of Palk Strait

The Palk Strait has the most typical change in marine environment around the country. Located between the southeast end of India and the northwest of Sri Lanka, the Palk Strait connects the Bay of Bengal and the Gulf of Mannar. 137 km long, 64–137 km wide, the Gulf of Mannar has an average depth of 2–3 m and a maximum depth of about 9 m. Large ships can only circle along the sea off Sri Lanka island. The Palk Bay, the southwest of the Bay, is also injected by Vaigai River and other rivers. The Strait serves as a bridge for trade between Jaffna port, a commercial center in the north of Sri Lanka, and southeast India (Fig. 164).

Except for the northwest of the Palk Strait, the main island has a short continental shelf with limited river input. Therefore, the scope of case 1 water in the offshore is very narrow and most of them are clean ocean. The Palk Strait in the northwest of the country is obviously affected by the input of suspended sediment from land-based rivers and the resuspension of sediments due to the shallow water depth. The water is relatively turbid, especially with the Calimere of India as the source of turbid zone. The Palk Strait and surrounding waters are the most typical case 2 water dominated by suspended solids in Sri Lanka, with the largest spatial-temporal change in water color in the fourth quarter of recent years from December to February. There are often reports of algae outbreaks with its important impact on local aquaculture (http://www.eosnap.com/tag/park-strait/page/3/). As the two ends of the Strait is the Indian Ocean affected by the joint action of wind, land-based rivers and ocean water bodies, and the spatial-temporal distribution of the bio-optical characteristics of the waters in the Strait is wide. Thus, the seawater is closely related to the economic development, leisure life and food health of people on both sides of the Strait. The spectral maximum value of MERIS remote sensing reflectance (2003–2012) acquired

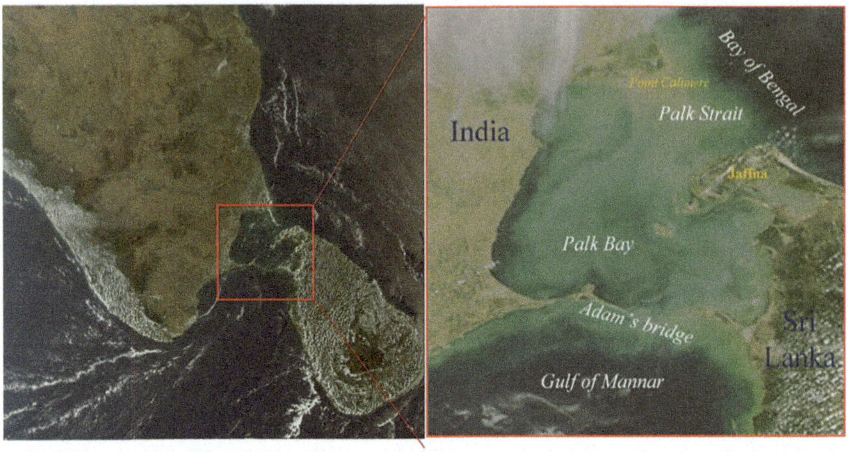

Fig. 164 MODIS Remote Sensing Image of Palk Strait (left: February 28, 2009, right: June 8, 2009)

through supervised classification algorithm can be applied for understanding the bio-optical characteristics of the high spatio-temporal distribution of the Palk Strait and its surrounding waters in different periods and places to provide the spatial distribution of the coastal conditions and understand the important ecological and biogeochemical characteristics of the upper ocean shelf.

Analysis on the Spatial-Temporal Distribution of Waters in Palk Strait

The Palk Strait and its surrounding waters are located on the continental shelf, whose color is mainly composed of suspended particles imported from land sources and resuspended sediment (with similar optical characteristics as the Yellow Sea). The following is the remote sensing reflectance spectrum classification and spatio-temporal characteristics analysis of the Strait based on the spectral classification method of water bodies in the Yellow Sea. The correction is as follows (Fig. 165).

The normalized spectrum and bio optical characteristics of Yellow Sea waters from class A to class E are as follows.

Class A: eutrophic water, characterized by the peak value of the R_{rs} spectrum curve in the visible light band at 400 nm (or <412 nm). The reflectivity rapidly

Fig. 165 Five different wates

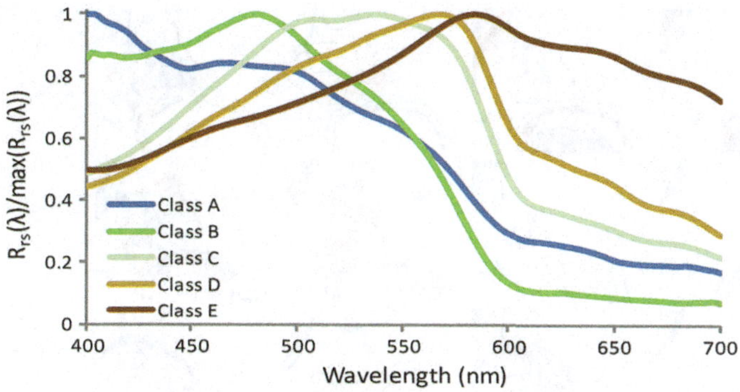

Fig. 166 Curve of spectral reflectance maximum of five waters by remote sensing

decreases with the wavelength and tends to 0 if it is more than 600 nm. According to the statistics of the Yellow Sea in China, the water generally is similar to the ocean Case 1 water, far away from the mainland and river thus rarely affected by land-based materials. Its color is sky blue, the FU value range is 4–7, the transparency is very high and the median depth of the Secchi disc is 12.0 m. The content of inorganic suspended sediment is as low (<1 mg/L) as the yellow matter. However, the concentration of chlorophyll dominates spectral change despite that it is relatively low. It is the transition between case 1 water and case 2 water (Fig. 166).

Class B: it is a common dominant type of chlorophyll and yellow matter. The peak value is near 490 nm and drops sharply to about zero near 600 nm, and most of them have absorption peak near 443 nm. The change of spectral shape is mainly affected by chlorophyll and yellow matter, with low suspended sediment in the deeper waters far away from the shore. The water color is greenish and blue, the FU value range is 5–11, the transparency is good, and the median depth of Secchi disc is 9.0 m. It is the most common type in clear water area in the offshore.

Class C: it is a mixed transition type and the peak of the water body's R_{rs} spectrum changes smoothly between 500 and 550 nm. The concentration of suspended sediment is lower (1.5–3 mg/L), but has a significant effect on the spectral shape due to the backscatter effect. It is greenish yellow, the range of FU is 6–14, the transparency of water body is medium, and the median depth of Secchi disc is 6.0 m.

Class D: it is dominated by suspended sediment. The minimum value of the R_{rs} spectrum is 40 nm, and the maximum value is near 565 nm. It is dominated by the backscattering of suspended solids. The spectrum increases monotonously with the wavelength first and then decreases with an obvious absorption inflection point near 600 nm. The water color is yellow, FU value range is 7–18, water transparency is poor, and the median depth of Secchi disc is 2.8 m. The water is affected by the re suspended particles or river input and the concentration of suspended solids is about 10 mg/L, which is mostly distributed in the near shore or shallow sea.

Class E: super high suspended sediment type. The minimum value appears at 400 nm and the maximum value appears after 570 nm with the 400–565 nm spectrum

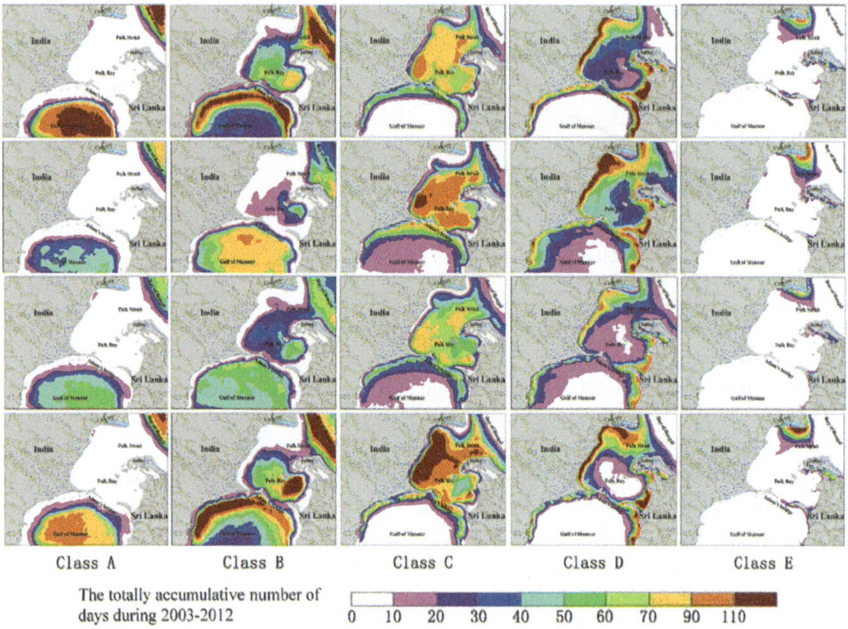

Fig. 167 Frequency chart of total cumulative distribution of waters in different seasons from 2003 to 2012 (from the first quarter to the fourth quarter from top to bottom)

almost in a straight line. The change of the remote sensing reflectance is mainly reflected at 560–700 nm or longer, distributed in the shoal and estuary area. The water color shows brown or yellowish brown color with the FU value range is 18–21 and the water transparency is extremely poor with the median depth of the cork disk is 0.4 m. The concentration of suspended solids is usually close to or greater than 100 mg/L (Fig. 167).

It can be seen that different waters vary greatly in time and space in terms of distribution range and intensity in different seasons.

In terms of spatial distribution, Manar Bay and Bengal Bay are far away from the sea or deep-water area (such that the water depth is more than 100 m), mainly class A and class B dominated by chlorophyll and yellow substances in poor nutrient water body. That means the water is relatively clean, transparent (usually more than 12 m), and the water color is sky blue, less affected by land-based materials and low particulate matter. With less particles (<10 mg/L), it has the characteristics of open ocean water (Indian Ocean) i.e., the dominant substance in the water is chlorophyll concentration and other substances that can be replaced by chlorophyll concentration or ignored. There is a narrow eutrophic type II water zone in the near shore shallow water area. The nearer the shore, the more turbid it is, the color of the water begins to change from blue to green yellow. In the southeast area of Delft Island in the central part of the Palk Strait, the water is relatively clean compared with others. The water is mainly class B and transitional class C with great changes throughout the year.

Nevertheless, the turbid suspended sand dominates the northwest of class D and the offshore area of the Palk Strait near Jaffna is also turbid all the year with class D due to the shallow water depth. The west and north of the Palk Strait is part of the delta of the Cauvery River, mostly mouths and shoals. It is turbid all year round and the water is yellow brown. The class E corresponding to the ultra-high suspended sediment dominates the water, the concentration of suspended sediment is close to or greater than 100 mg/L and the transparency of the water body is poor. With the Calimere as the source of turbidity, it is mainly because the Cauvery River transport a large amount of sediment. As the coastal river enters into the Strait, the water color changes with the strength of the sediment. Due to the shallow depth, Jaffna lagoon is turbid all year round, mainly class D and class E. In addition, the Adam's Bridge, may also be occasionally turbid due to the shallow water.

In terms of time distribution, the waters of Manar Bay and Bengal Bay are relatively clear in the fourth and first quarter. Most of the sea are mainly Class A of blue ocean eutrophic water and the coastal waters are mainly Class B of phytoplankton rich. In the second and third quarter, the water is blue-green as the concentration of particulate matter and phytoplankton in the water increase. The sea is dominated by Class A of ocean eutrophic water and Class B of phytoplankton rich water. Especially in the second quarter, class B of phytoplankton rich water is the main part of Manar Bay, so it becomes blue-green with more concentration of phytoplankton and other particles. In the first and fourth quarter, the water of the Palk is relatively clear, especially blue-green of class B dominated by phytoplanktons in the southeast corner of Delft Island. However, the water becomes turbid in the second quarter with transitional class C and class D dominated by suspended solids, and the transparency became the worst once a year. Both the Strait and the Lagoon are dominated by suspended solids with high concentration, with similarly Class E optical properties. The coverage is wider in the second quarter, while in the third quarter, the frequency and coverage are the least.

Remote Sensing Analysis of Socioeconomic Development

1 Economy

1.1 Economic Profile

Sri Lanka is an agricultural country with a plantation economy, producing tea, rubber, coconut and rice. It prospers the agricultural products and garment processing industries but is weak in the industrial sector. Under the economic liberalization policy in recent years, its economy has maintained a medium growth rate and a good momentum of development.

Agriculture: tea, rubber and coconut are the three sources of Sri Lanka's agricultural revenue. The proportion of agricultural production in GDP has been declining from 50% in the 1950s, 35% in the 1970s, 20% in the 1990s, to 10% in 2014 due to the high agricultural production cost, low productivity, and large losses as well as the rising non-agricultural outputs.

As an important part of the national export earnings, agricultural exports accounts for about 25% of foreign exchange earnings. Tea, rubber and coconut are the three major commodities of the agricultural exports and tea exports occupy 59% of the total agricultural products.

Industry: Sri Lanka is relatively weak in its industrial sectors. A large number of industrial raw materials still need to be imported from abroad due to the lack of resources hence no capital-technology-intensive industry and almost no heavy industries. Nowadays, the industries mainly produces textile, clothing, leather, food, beverage, tobacco, chemicals, petroleum, rubber, plastic and includes non-metallic mineral products processing industry, mining and quarrying industry. In 2014, the industrial sector accounted for 34% of the national economy.

Textile and garment industry are the pillar industry of the national economy and the largest export-oriented industry, taking up 43% of the national foreign trade exports. In addition, Sri Lanka is famous for its gemstones and their processing that accounts for 4% of the total exports.

X. Wei, *The Geography and Remote Sensing Analysis of Sri Lanka*,
https://doi.org/10.1007/978-981-33-4627-7_3

Service industry: it consists of wholesale and retail, hotel, catering, transportation, warehousing, information and communication, tourism, financial services, real estate and commercial services, public management and other social and personal services.

Recently, the government are making efforts to transform into a service-oriented economy. Currently, the service industry has become its leading industry and the main driving force of its economic growth, especially the growing information and communication industry. In 2014, the production of service industry accounted for 56% of GDP.

Foreign trade: it adopts a free foreign trade policy, which means that all goods can be imported freely except for the oil controlled by the government. In recent years, the structure of export trade has changed completely from agricultural products to industrial products. At present, the main export products are textiles, clothing, tea, rubber and its products, jewelry products, etc. The main importer are the United States, the United Kingdom, India, Italy and Germany and the main exporter are India, China, the United Arab Emirates, Singapore and Japan.

Source: website of the Economic and Commercial Counsellor's office of the Chinese Embassy in Sri Lanka.

1.2 Spatio-temporal Analysis of Sri Lanka Economy Based on Nighttime Light Data

DMSP-OLS nighttime light data widely applied in urban, population, economic development and other fields is a good tool for monitoring human activities and effects.

DMSP-OLS data is obtained by the Operational Linescan System (OLS) on the Defense Meteorological Satellite Program (DMSP) launched by the National Oceanic and Atmospheric Administration (NOAA). The DMSP satellite system (F10, F12, F14, F15, F16, F18) launched by NOAA has been equipped with OLS sensors since 1992, which can work at night and detect the low-intensity night lights emitted by urban night lights or even small-scale residential areas and traffic flow and make it different from the dark countryside. Therefore, DMSP-OLS night light image can show human activities and serve as a stable data source for human activity monitoring research.

DMSP-OLS data includes Average Visible, Cloud Free Coverage, and Web Stable Lights. The Web Stable Lights with the spatial resolution of 890 m × 890 m reduces the errors such as lightning and fire light and improves the analysis accuracy. The development of the urban economy will be analyzed combined with the total nighttime light (TNL) of the Web Stable Lights and the national GDP.

The relationship between the light intensity and the GDP has been analyzed through correlation analysis and regression analysis based on the national DMSP/OLS NTL image from 1992 to 2012 and the national GDP issued by the World Bank. It shows that there is a significant correlation between the DMSP-OLS

NTL data and GDP and the correlation coefficient reaches 0.771. Moreover, the correlation coefficient is calculated as 0.97 (significant correlation if P is 0.01) based on the national GDP and gray-scale (DN) value of DMSP-OLS NTL from 1992 to 2012, which means the total figures are highly correlated and DMSP/OLS NTL can be used as data evidence. Given this, the national GDP grid image at the scale of 1000 m × 1000 m can be constructed. The grid image demonstrates GDP by the geographic location with high frequency of human night activities highlighted by the light and shadow image except agricultural figure (Fig. 1).

The following grid chart shows an increasingly growing national economy from 1992 to 2012. Specifically, the economy of the southwest coastal region that has developed earlier is now gradually benefiting the inland. The northern, eastern and southeast parts are still marginalized in the economic growth. Colombo and its surrounding regions have witnessed the highest growth rate as it is the capital of Sri Lanka and one of the main trading port. Given its strong economic base, it now enjoys higher economic density and a much higher growth rate than before (Fig. 2).

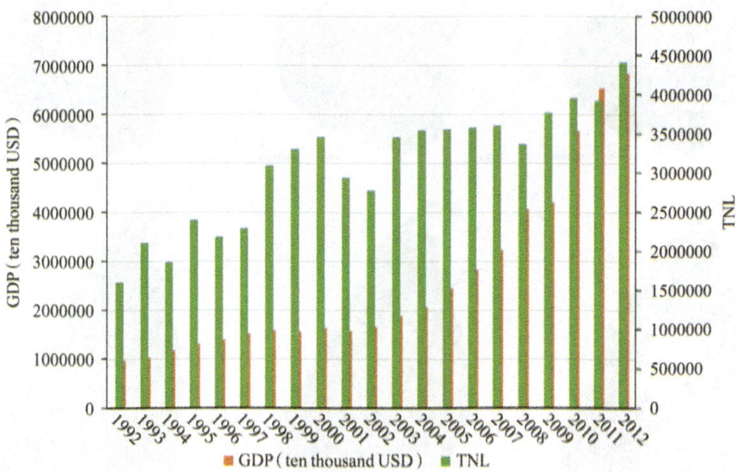

Fig. 1 GDP and TNL in 1992–2012

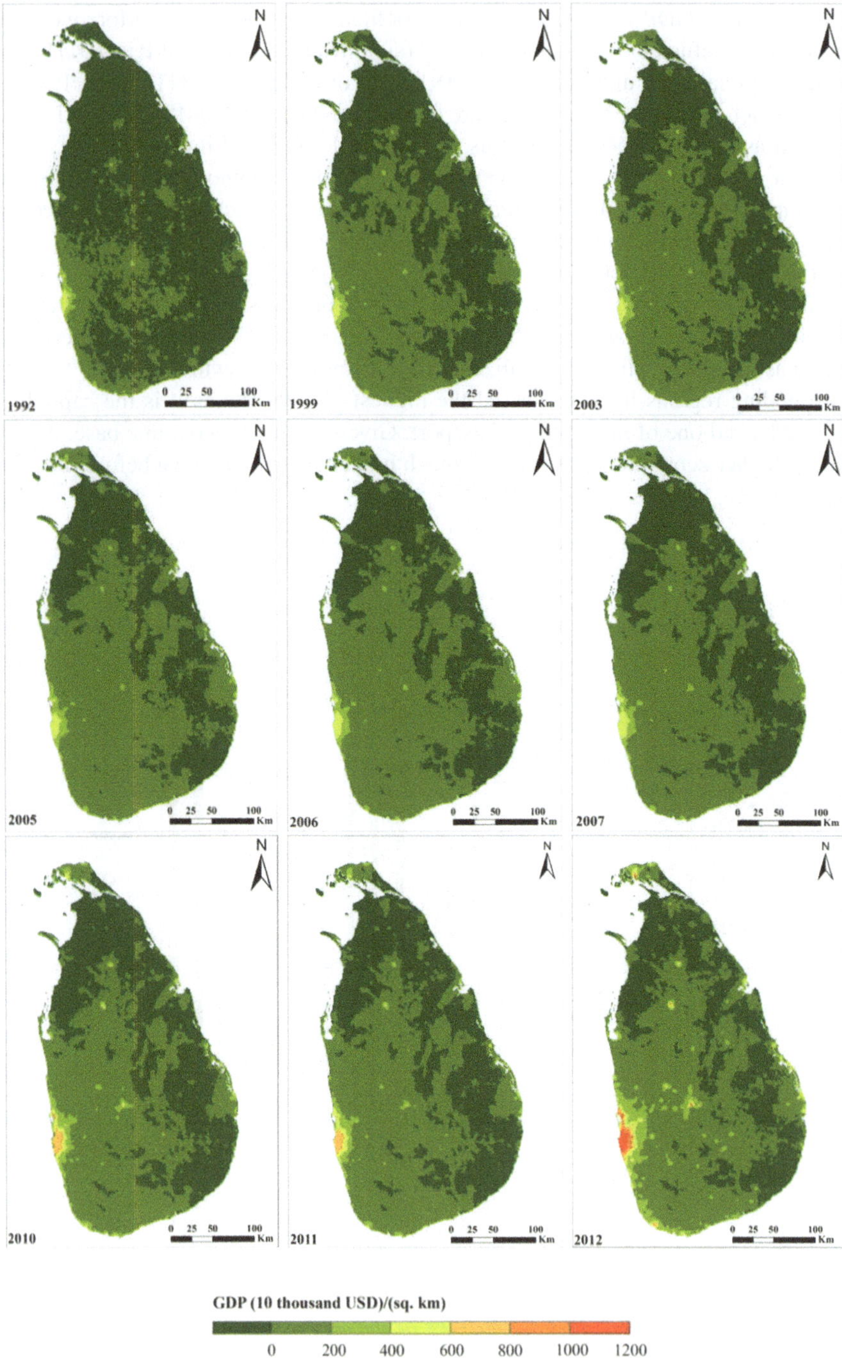

Fig. 2 Grid image of GDP from 1992 to 2012

Table 1 National roads (class A, class B, class E)

Road class	Length (km)
Class "E" roads	**169.8**
Class "A" roads	**4217.4**
Class "AA" roads	3720.3
Class "AB" roads	466.9
Class "AC" roads	30.2
Class "B" roads	**8003.2**
Total of "A" and "B" class roads in Sri Lanka	**12,220.6**
Grand total of national highways in Sri Lanka ("A", "B" and "E" class roads)	**12,390.4**

2 Transportation

The national traffic is mainly based on the road network centered in the capital with few railway transportations lines. With the recovery of global economy, the port construction and transportation show a significant growth trend. On the other hand, the country makes efforts to develop the air transport industry.

2.1 Roads

Roads, occupying more than 93% of the national land transportation, serve as the main way of transportation. Admittedly, Sri Lanka is one of the countries with the highest road density in the world. The national roads are divided into class A and class B and the former class can be divided into class AA, class AB and class AC. Also, Provincial roads are divided into class C and class D. Currently, the total mileage of paved roads is more than 100,000 km with the road density of 1.5 km/km^2. The total mileage of national roads reaches 12,200 km and that of E-class roads, the newly-constructed road, is nearly 170 km.

The roads in the southwest region, especially around Colombo is in good condition with flat asphalt and clear signs and have the highest across the country. In addition, busy roads are being upgraded (Table 1 and Fig. 3).

2.2 Railways

The rail transport includes an inter-city network connecting major cities and commuter rail for Colombo.

Composed of main line, coastal line and inland line with Colombo as the center, the railway network connects all the main cities. The total mileage of the national

Fig. 3 Road network. *Source* Road Development Authority, Sri Lanka (July 31, 2018)

Fig. 4 Coastline railways: Colombo to Galle. *Source* Department of Railways, Sri Lanka

railways was 1,568 km by the end of 2017. With the end of the civil war, the country further accelerated the reconstruction and repair of the northern railways (Figs. 4 and 5).

2.3 Airports

International airports: Sri Lanka has three international airports, namely, Bandaranaike International Airport, Mattala Rajapaksa International Airport and Ratmalana International Airport under reconstruction. Bandaranaike International Airport, located 35 km north of Colombo, is the hub airport of the National Airlines and the largest international airport in the country, with flights reaching all over the world. Mattala Rajapaksa International Airport, located in Hambantota, is the second international airport. It started operation in October 2012, officially opened in 2013 and closed in June 2018. Ratmalana International Airport, located 15 km south of Colombo, is the first international airport and was the only international airport before the completion of Bandaranaike International Airport in 1967. After the opening of Bandaranaike, Ratmalana transferred all international services to the new airport. In 2012, Bandaranaike announced the opening of small international commercial flights, mainly serving private and small aircrafts.

Domestic airports: Sri Lanka has 19 domestic airports, 15 of which are paved with runways while 4 are not (Fig. 6).

Fig. 5 Railway network

Fig. 6 Main airports

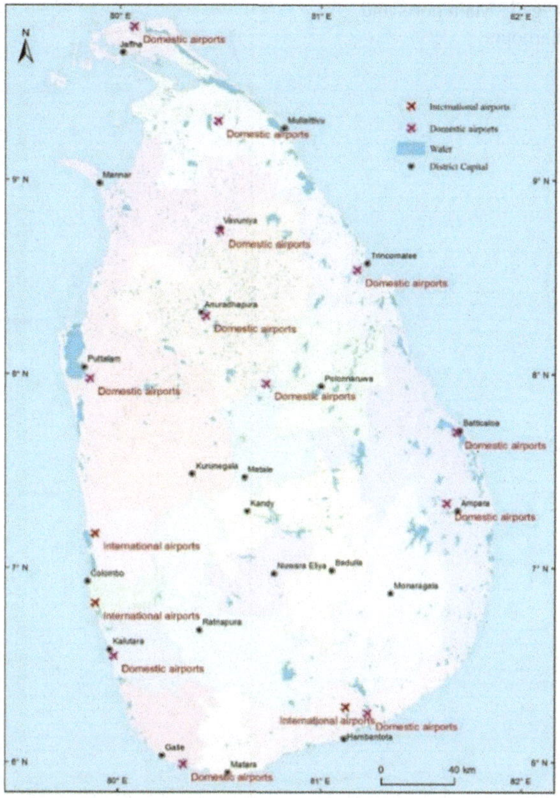

2.4 *Ports and Harbours*

As an island country at the center of the Indian Ocean route. Sri Lanka has many natural ports due to its tortuous coastline. Since the opening of the new route, it has been playing an irreplaceable role in maritime transport.

The main ports are Colombo, Galle, Hambantota and Trincomalee (Fig. 7 and Table 2).

Colombo Port

Colombo Port bordering on the north of the Indian Ocean, located in the south bank of Keleni estuary on the southwest coast of the country, is the largest port in the country. With railways and roads across the country, it serves as the national transportation hub. The Port, also known as Colombo Artificial Port, is one of the largest artificial ports in the world and one of the important midway ports of the world's navigation lines in Eurasia, the Pacific Ocean and the Indian Ocean.

It was known to Roman, Arab and Chinese businessmen more than 2000 years ago and has developed into an important trading town in the 8th century A.D. The modern Colombo Port was built in the 19th century during the British colonial period and it

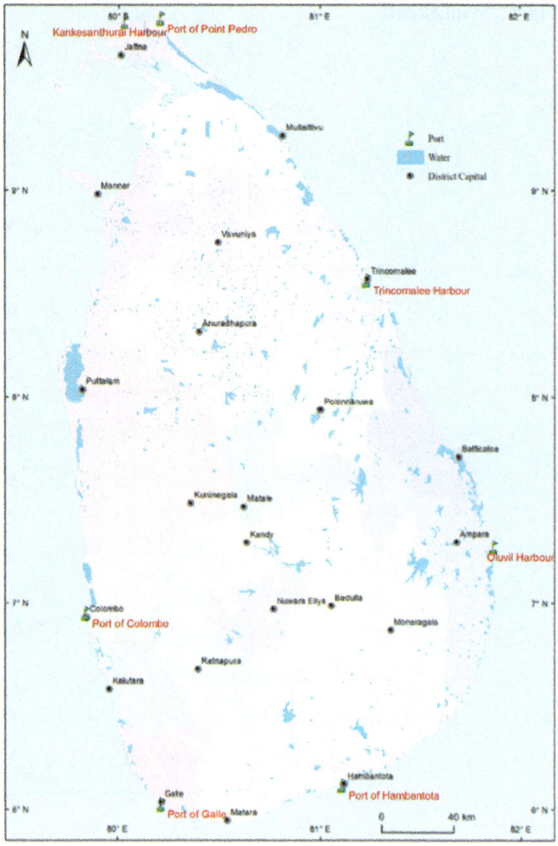

Fig. 7 Main ports and harbours

Table 2 Main ports and harbours

Name	City	Description
Port of Colombo	Colombo	The largest in the country
Port of Hambantota	Hambantota	New in the south
Trincomalee Harbour	Trincomalee	The fifth largest natural port
Port of Galle	Galle	Regional port
Port of Point Pedro	Point Pedro	A small port at the north end of the country
Kankesanthurai Harbour	Kankesanthurai	A small military port
Oluvil H Oluvil Harbour	Oluvil	Newly open in 2013

has been rapidly modernized with cranes, gantry cranes and other modern terminal facilities since the 1980s.

Colombo Port Expansion Project (CPEP) was started in 2008 and Colombo South Container Terminal was launched in 2011 and put into operation in August 2013. The eastern port was completed and put into operation in 2017.

Colombo Port City

It is located in the core of Colombo CBD, close to the landmark buildings such as Hilton Hotel and known as the future city. Jointly developed by China and Sri Lanka, it has the planned construction scale of more than 5.3 million square meters including 276 ha of land reclamation planned to be completed in three years. It is to generally complete in five to eight years and fully in 20 to 25 years. The project was started on September 17, 2014, suspended on March 5, 2015 and resumed in September 2016.

The following figure shows the satellite remote sensing images of Colombo Port and Colombo Port City in different periods with high spatial resolution from Google Earth, directly showing whole project progress (Fig. 8).

Hambantota Port

Also known as Hambantota Deep Water Port, it is located on the southern coast of Sri Lanka 165 km away from Colombo. The Eastern Crossroads serves as a gold position within 10 nautical miles from the main channel between Asia and Europe and an important port along the Belt and Road.

The Hambantota Port Development Project (HPDP) includes the comprehensive hub port and the surrounding industrial area. The first phase of the project was started in January 2008, completed in December 2011 and put into operation from June 2012. The second phase was started in September 2012 and completed in April 2015. Hambantota Port is a full deep-water port with 10 berths in the first and second phases, a length of 3487 m. The berths specially serve for containers, bulk cargo, general cargo and liquid bulk cargo. The water depth of the wharf and navigation channel is 17 m, enough for super large ships.

Figure 9 shows the high-resolution remote sensing images of different periods around Hambantota Port, directly showing the project progress of infrastructure construction.

3 Major Cities

The city is the settlement where the society develops to a certain stage, serving as the most active area during social and economic development.

Sri Lanka's urban layout differs from that of China due to their differences in geography, natural climate and land policy. China's cities are mainly concentrated in blocks, while Sri Lanka's cities are mostly distributed in strips due to natural conditions or traffic lines along the coastline, rivers, and long and narrow valleys,

Fig. 8 Time series remote sensing image of Colombo Port and Colombo Port City

Fig. 8 (continued)

hence elongated directional cities with strong direction of plane structure and traffic flow.

There are obvious spatial differences in urban and rural distribution structure in mountainous areas and plains, in north and south. In the plain area of the southwest, towns and villages have no obvious boundaries. Either in the city or in the countryside, the houses are surrounded by tropical palms and exotic plants, forming a beautiful scenery and elegant environment. However, due to the adverse effects of civil war, the water conservancy facilities destruction and the large number of people moving southward from the north, the villages are now relatively sparse, small in scale and relatively simple in housing, which are often regarded as remote areas.

Fig. 8 (continued)

At present, Sri Lanka is witnessing its rapid urbanization. As an important indicator of urbanization process, the spatial expansion of urban built-up land is the result of many factors such as nature, economy, policy and so on. Through the analysis and research of urban expansion, the growth process and evolution law of the city can be revealed to deepen the understanding of the essence of urbanization and provide effective spatial decision-making information for urban development planning and regulation.

Based on Landsat-TM images of 1995 and 2005 and Landsat 8-OLI images of 2015 by the remote sensing monitoring of spatial expansion process of main urban built-up areas, the characteristics and impact on surrounding land use in the process

Fig. 8 (continued)

of expansion can be analyzed in case for the references for study and decision-making for its urban development. The main monitoring objects include four important port cities, namely Colombo, Galle, Hambantota and Trincomalee, as well as three famous ancient cities, Anuradhapura, Polonnaruwa and Kandy (Fig. 10).

Fig. 8 (continued)

3.1 Colombo

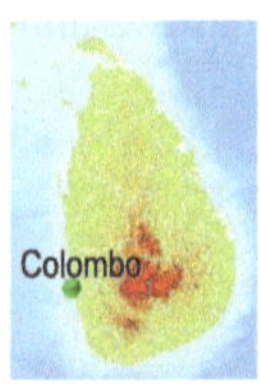

Fig. 8 (continued)

Colombo, the capital of Sri Lanka, is its political, commercial and cultural center. Located on the southwest Bank of Sri Lanka Island, close to the Indian Ocean, it is an important port in the Indian Ocean, known as the Asian gateway and Oriental crossroads. As the largest city nationwide, Colombo has a population of 752,900 and its whole administration area has a population of approximately 2,323,800 from 2012 census. Sinhalese are the majority and Tamils, Moors, Indians, Malays are the rest. Buddhism dominates, followed by Islam, Christianity and Hinduism. Fort District in the north is the center for government agencies, banks, markets, tourism, aviation and steamship companies (Fig. 11).

History: Colombo City has a long history. It served as a Muslim commercial town around 700 A.D. where Arabs settled here and named it Kalanpu. In the 16th century, Portuguese colonists established their military fortresses here and renamed it. It has been the main foreign trade port since the 17th century for shipping between Europe, Asia, Africa and the Pacific. It became the capital after the national independence.

Climate: Colombo is located near the equator with a tropical monsoon climate. The annual average temperature is 22–32 °C and the annual average rainfall is about 2,300 mm. In most cases, the climate is mild and humid with abundant rain, the scenery is pleasant with lush trees.

Transportation hub: a transportation hub for East-West shipping in the Indian Ocean, Colombo Port serves for more than 90% of national imports and exports. Bandaranaike International Airport has flights to major cities all over the world. Inland transportation is convenient with the railways and highways across the country.

Urban economy: its traditional economy is mainly agriculture, port industry and service industry. Most of the emerging manufacturing factories are built on the outskirts of the city. Although the industry is underdeveloped, there are industries such as textile, tobacco, machinery, metal, food, chemical engineering, radio, oil

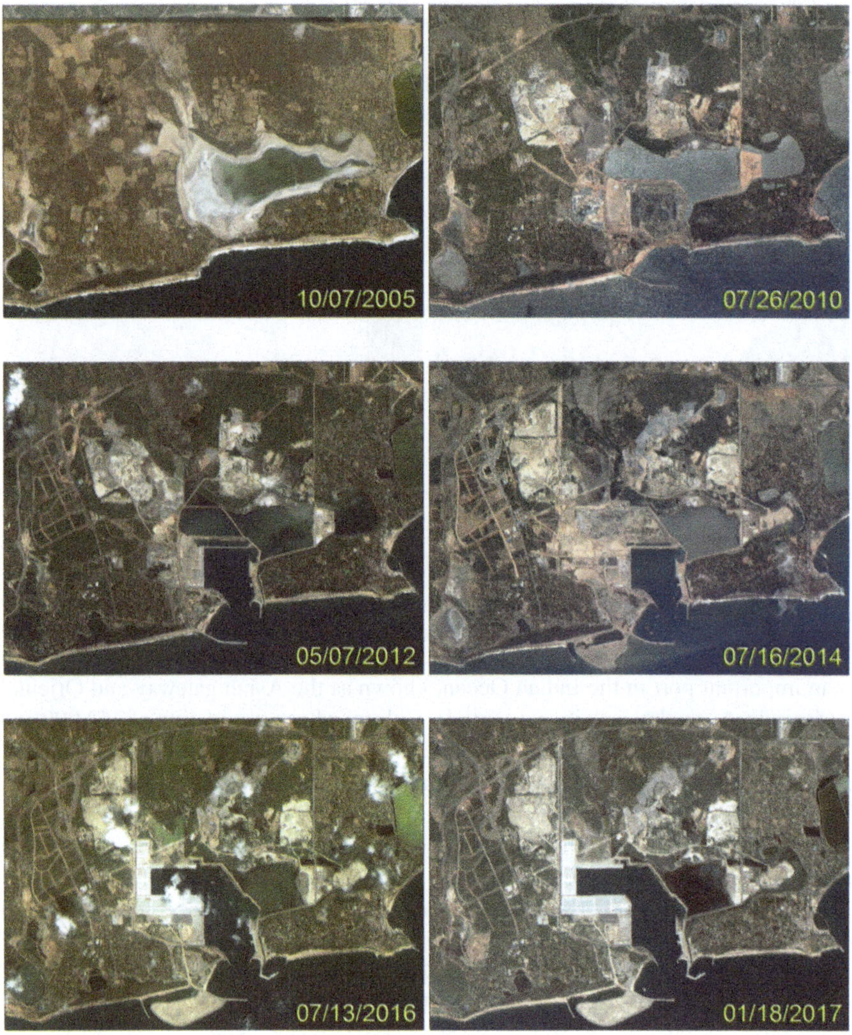

Fig. 9 Remote sensing image of Hambantota port

refining, chemical fertilizer as well as general machinery and automobile assembly and repair plants. There are 4 hydropower stations in the upper reaches of the Kelani River, which supply power for the whole city.

Cultural Center: there are many famous public schools. Most famous schools in Colombo can be traced back to the 19th century, such as the Royal College Colombo built in 1835. There are also many places of interest. Colombo Museum, built in 1877 on Fernando Avenue, is the oldest museum in Sri Lanka, which collects historical relics of various periods. Dehiwala Zoological Garden and Viharamahadevi Park

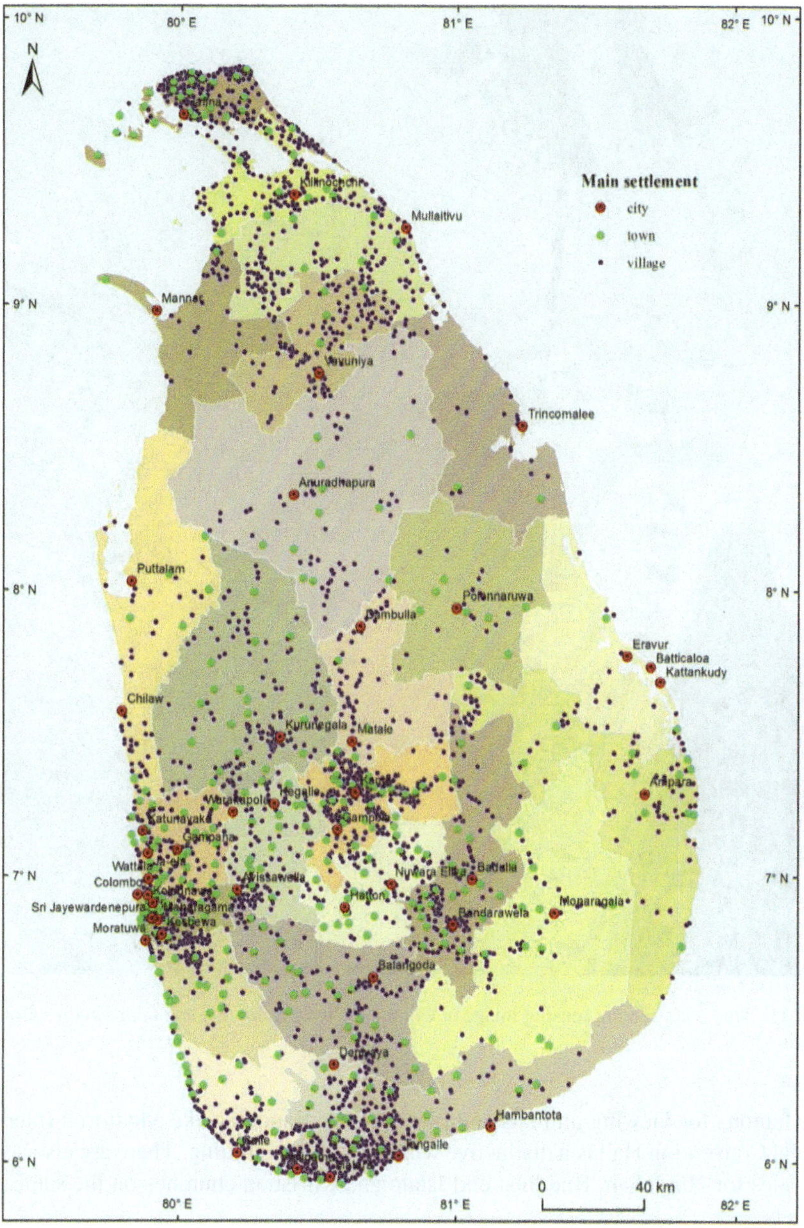

Fig. 10 Main towns and residential sites

Fig. 11 True color remote sensing image of Colombo City. *Remote sensing image source* Google Earth

are famous for viewing animals and plants. Also, Bandaranaike Memorial International Convention Hall is a distinctive white octagonal building. There are also some temples for Hinduism, Buddhist and Islam and Christian churches on the same old streets.

Remote Sensing Monitoring of Colombo Urban Expansion from 1995 to 2015
Characteristics of urban expansion in Colombo: the built-up area increased by nearly half from 1995 to 2015, making itself to be the city with the national largest increase in built-up area at that time. There are significant differences in the number and space

of Colombo urban expansion between the period of 1995–2005 and that of 2005–2015. In terms of the quantity, the built-up expansion area increased from 680.90 km^2 in 1995, 862.17 km^2 in 2005 with a net increase area of 181.27–997.69 km^2 in 2015 with a net increase area of 135.52 km^2. The later expansion area is smaller than that earlier with the average annual expansion rate reducing from 2.39 to 1.47%. In light of the spatial expansion distribution, the urban development focuses on the increase of internal filling and building density, supplemented by edge expansion. Periodically, the progress is divided into the eastern and southern part in the first decade and the northeastern part in the second decade (Fig. 12, Table 3 and Fig. 13).

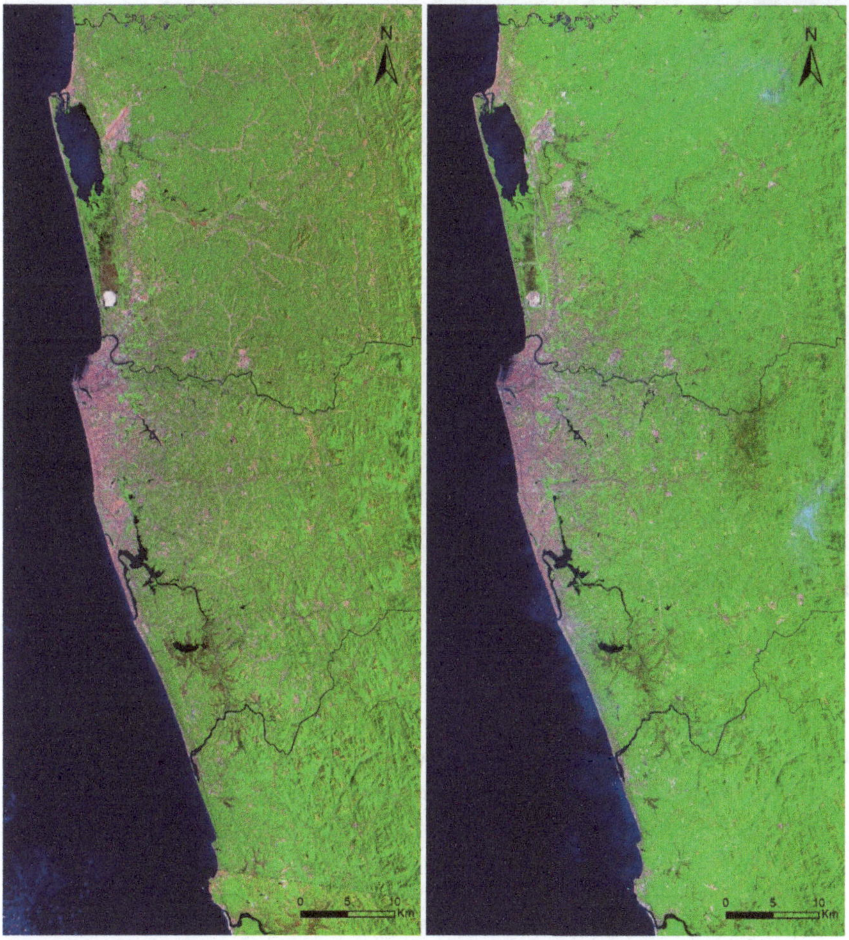

(a) Colombo remote sensing image in 1995 (b) Colombo remote sensing image in 2005

Fig. 12 Remote sensing image by years and built-up expansion image. The yellow part in the Figure **d** indicates the built-up areas in 1995, the green for the urban expansion area in 1995–2005 and the red for the urban expansion area in 2005–2015

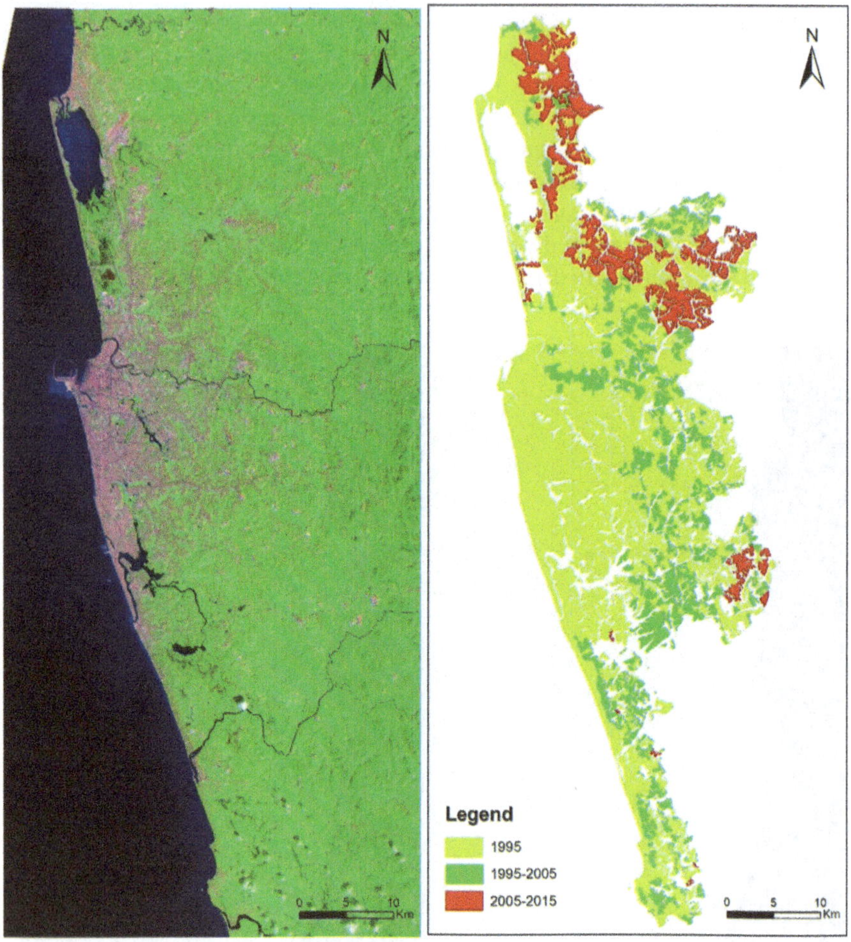

(c) Colombo remote sensing image in 2015 (d) expansion of Colombo built-up areas from 1995 to 2015

Fig. 12 (continued)

Table 3 Expansion quantity characteristics of built-up area

Year	Area (km^2)	Expansion area (km^2)	Expansion speed (km^2/a)	Annual average growth rate (%)	Expansion density
1995	680.90				
2005	862.17	181.27	18.13	2.39	2.66
2015	997.69	135.52	13.55	1.47	1.57

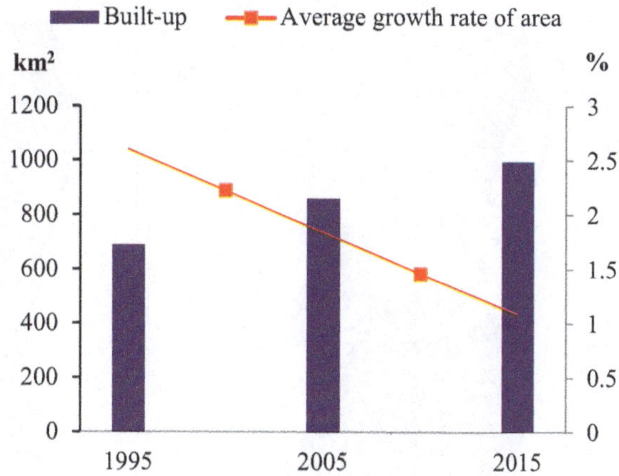

Fig. 13 Built-up area and annual growth rate in 1995, 2005 and 2015

The occupied land during the urban expansion includes forest land of 246.64 km^2 and 70.15 of cropland in the past two decades. This is because the residential land is often surrounded by luxuriant trees. The priority of the urban expansion is to occupy the nearby forest land, causing less forests and more buildings. There are differences in the number and distribution of occupied land in the past decade. From 1995 to 2005, 161.18 km^2 of forest land and 20.09 km^2 of arable land were occupied, which are 18.69% and 2.33% respectively of the urban built-up land in 1995. The occupied arable land is mostly located in the southeast edge, while the forest land, 7 times larger than the cropland, is scattered in the surroundings of the internal buildings on the existing built-up land. From 2005 to 2015, the occupied area of forest land and cropland is 85.46 km^2 and 50.06 km^2 respectively, or 8.57% and 5.02% of the urban built-up land in 2005. After these ten years, there were few forest lands available for development in the built-up land, so the newly one is mainly distributed in the northeast edge area (Figs. 14 and 15).

3.2 *Galle*

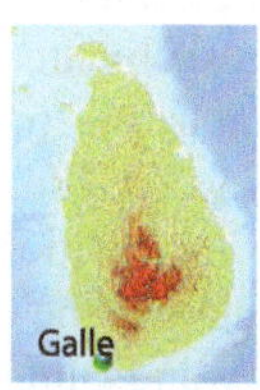

Galle is a port city and the capital of Southern Province, 115 km away from Colombo. The total population of Galle was 99,500 in 2011, of which more than 70% are

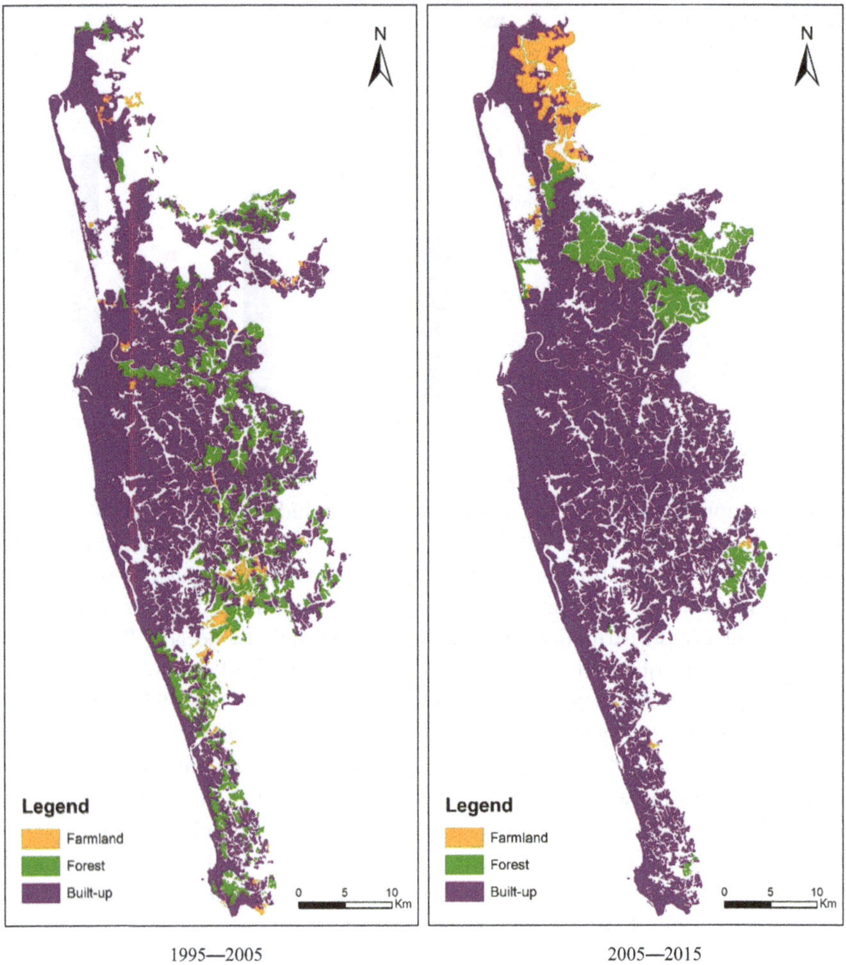

Fig. 14 Types of occupied land in urban expansion

Sinhalese. Traced back to the Solomon Dynasty, Galle is a city with a long history and one of the four national economic centers (Fig. 16).

Climate: Galle has tropical rainforest climate with the annual temperature changes of 2403 mm and the average temperature of 26 °C.

Architectural style: the Old Town of Galle fully shows the influence between European architecture and South Asian traditional architecture during the 17th and 19th centuries, as local architects integrated the original European architectural style with local geography, climate, history and culture.

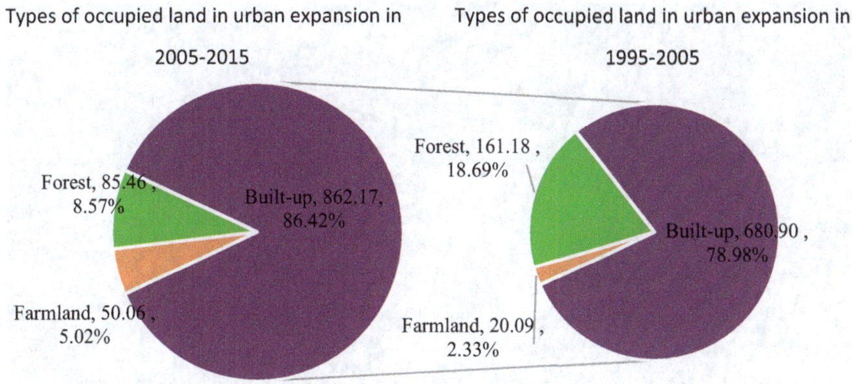

Fig. 15 Proportion of occupied land of new built-up land by periods

Culture: the Old Town of Galle, a UNESCO World Heritage site, is an old trading port with an exotic style where there are Dutch colonial buildings, ancient mosques and churches, magnificent buildings and museums.

Remote Sensing Monitoring of Galle Urban Expansion from 1995 to 2015
Characteristics of urban expansion in Galle: the built-up area increased by 19.57 km^2 from 1995 to 2015. Although the increased area is much smaller than that of Colombo, the increased rate is much higher, totaling 80% in the past two decades. There are differences in the number and space of Galle urban expansion between the period of 1995–2005 and that of 2005–2015. In terms of the quantity, the built-up expansion area increased from 24.71 km^2 in 1995, 36.41 km^2 in 2005 with a net increase area of 11.70–44.28 km^2 in 2015 with a net increase area of 7.87 km^2. The former yearly expansion rate was higher than that in the later period. In light of the spatial expansion distribution, the urban development focuses on the internal and southern expansion. Periodically, the progress is divided into the southern part in the first decade and then the internal and northern part (Fig. 17, Table 4 and Fig. 18).

Characteristics of occupied land during urban expansion in Galle: the land near the sea is not suitable for arable land, hence more farmland and forest land was occupied for development. For example, 26.92% and 5.22% of the total built-up area in 2005 were forest land and arable land respectively in 1995. Then in 2015, 16.60 and 1.17% of the total built-up area were those in 2005. Since tourism is the main local revenue, new built-up land has been built in the inland of the city in recent decade to protect local sites and develop tourism. The city expansion is temporarily stagnant due to its own geographical location, politics and economy (Figs. 19 and 20).

Fig. 16 True color remote sensing image of Galle City. *Remote sensing image source* Google Earth

(a) Galle remote sensing image in 1995

(b) Galle remote sensing image in 2005

(c) Galle remote sensing image in 2015

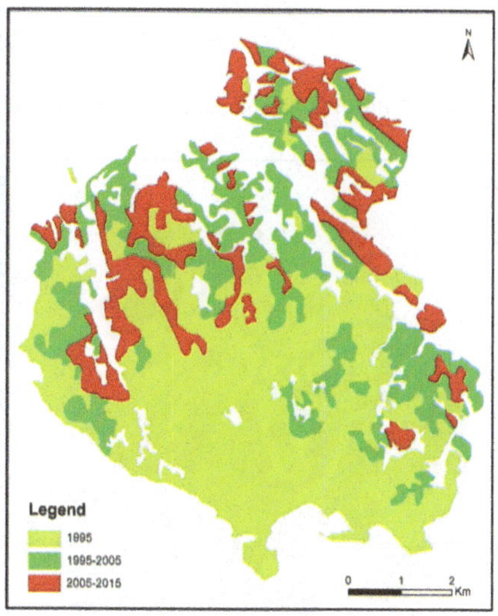

(d) built-up areas expansion of Galle from 1995 to 2015

Fig. 17 Remote sensing image by years and built-up expansion image in Galle. The yellow part in the Figure **d** indicates the built-up areas in 1995, the green for the urban expansion area in 1995–2005 and the red for the urban expansion area in 2005–2015

Table 4 Expansion quantity characteristics of built-up area in Galle

Year	Area (km^2)	Expansion area (km^2)	Expansion speed (km^2/a)	Annual average growth rate (%)	Expansion density
1995	24.71				
2005	36.41	11.70	1.17	3.95	4.74
2015	44.28	7.87	0.79	1.98	2.16

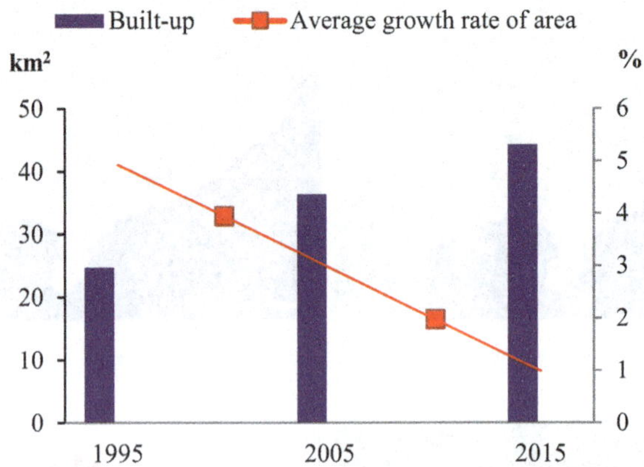

Fig. 18 Built-up area and annual growth rate of Galle in 1995, 2005 and 2015

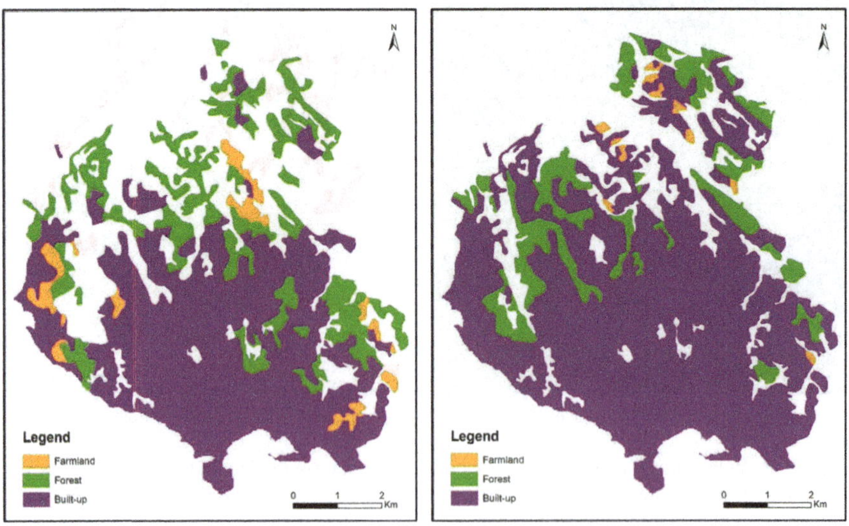

1995—2005 2005—2015

Fig. 19 Types of occupied land in urban expansion

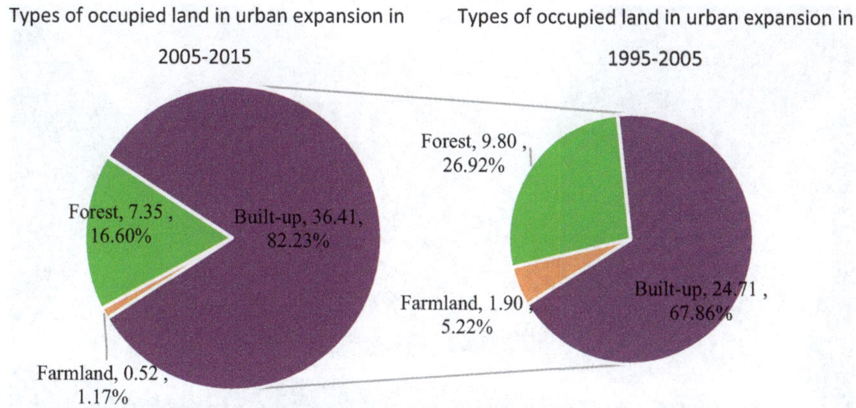

Fig. 20 Proportion of occupied land of new built-up land by periods

3.3 Hambantota

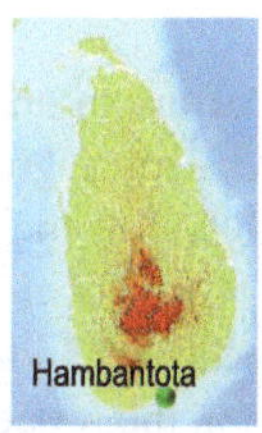

Located on the southeast coast, Hambantota is a port in the southeast of Sri Lanka, 237 km from Colombo. The majority are Malays who believe in Islam as well as some Sinhalas who believe in Buddhism (Fig. 21).

Climate: it is a typical tropical dry and wet climate. There is less rain from January to March and June to August and abundant rain in October and November. The yearly rainfall is about 1050 mm. The yearly average temperature ranging from 26 °C in January to 28.1 °C in both April and May.

Transportation: the highway connects Colombo and other places with A2 highway going through Colombo, Galle and Matara. Hambantota port is located in the southernmost part of South Asia, the only place for sea routes between the Middle East, Europe, Africa and the East Asia.

Remote Sensing Monitoring of Hambantota Urban Expansion from 1995 to 2015

Urban expansion characteristics of Hambantota: as a growing port city, Hambantota has made remarkable achievements in urban expansion in recent decade. The urban built-up area was 49.02 km^2, 62.79 km^2 and 99.89 km^2, respectively in 1995, 2005

Fig. 21 True color remote sensing image of Hambantota. *Remote sensing image source* Google Earth

and 2015. The average growth rate of urban area in the two periods was 2.49 and 4.75%. The urban expansion showed a rapid growth trend. The newly-built airport and the developing ports and railways with the support of China in recent decade have been the strong driving force for the local economic development and further rapid expansion of the city (Fig. 22, Table 5 and Fig. 23).

Characteristics of land occupation in the process of urban expansion: 18.12% and 9.78% of the total urban built-up land came from the forest land and cropland in 2005 and 1995 respectively. The forest land in the west of the city was mainly converted into built-up land, while the cropland in the east was mainly transferred into built-up land. 25.28% and 10.74% of the total urban built-up land in 2015 came from the forest land and cropland in 2005, with another 1.12% of the lakes occupied in 2005. In the past decade, the urban expansion focused on the construction of airports and related facilities in the north of the city and the construction of ports near the coastline. The government plans to convert Hambantota into a major center of the country, which will still accelerate the urban expansion in the future (Figs. 24 and 25).

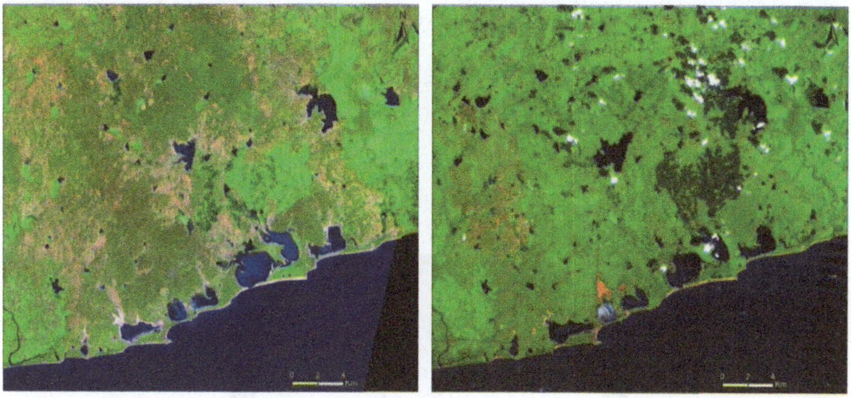

(a) Hambantota remote sensing image in 1995 (b) Hambantota remote sensing image in 2005

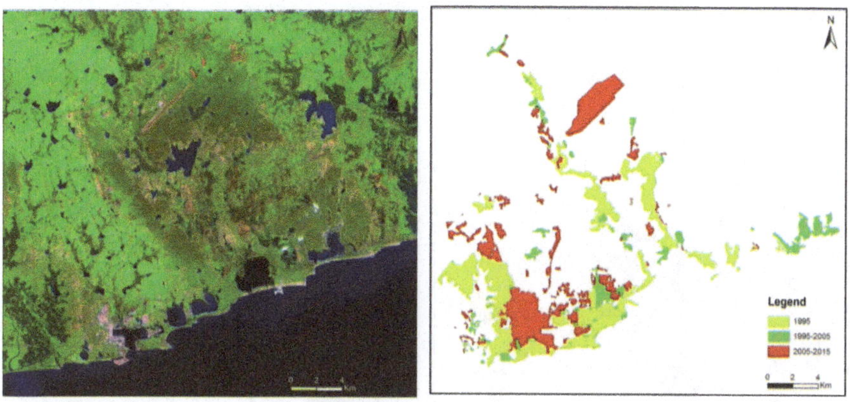

(c) Hambantota remote sensing image in 2015 (d) built-up areas expansion of Hambantota from 1995 to 2015

Fig. 22 Remote sensing image by years and built-up expansion image in Hambantota. The yellow part in the Figure **d** indicates the built-up areas in 1995, the green for the urban expansion area in 1995–2005 and the red for the urban expansion area in 2005–2015

Table 5 Expansion quantity characteristics of built-up area in Hambantota

Year	Area (km^2)	Expansion area (km^2)	Expansion speed (km^2/a)	Annual average growth rate (%)	Expansion density
1995	49.09				
2005	62.79	13.70	1.37	2.49	2.79
2015	99.89	37.10	3.71	4.75	5.91

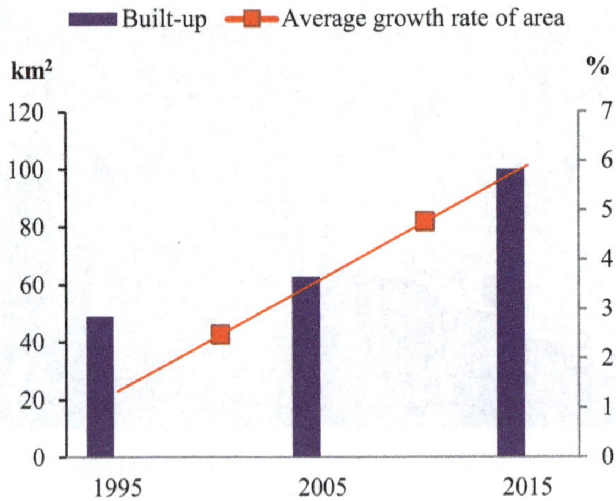

Fig. 23 Built-up area and annual growth rate in 1995, 2005 and 2015

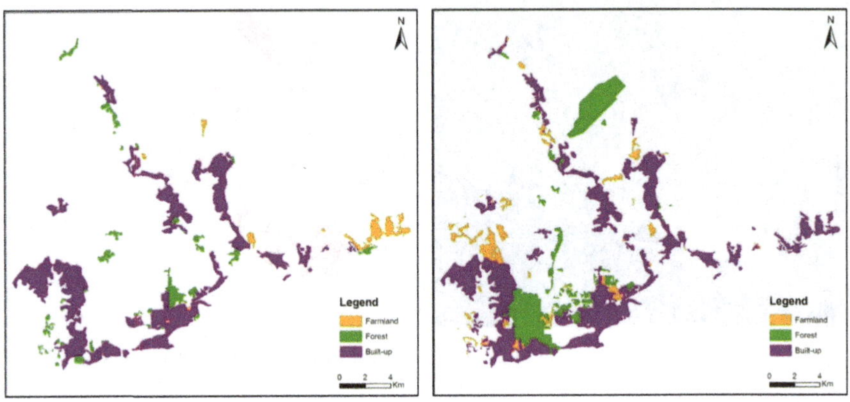

1995—2005 2005—2015

Fig. 24 Types of occupied land in urban expansion

3.4 Trincomalee

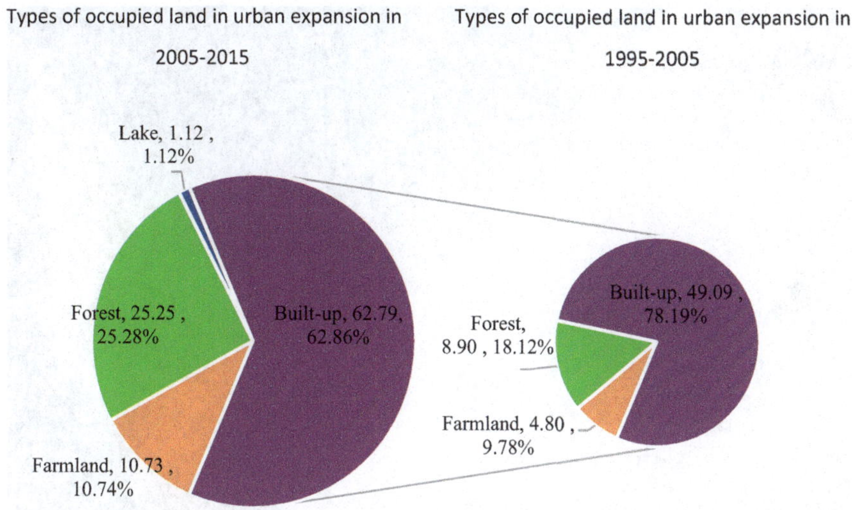

Fig. 25 Proportion of occupied land of new built-up land by periods

Located on the east coast and the north bank of Codiar Bay, Trincomalee Bay, Trincomalee is a port city with the national largest natural harbor and a beautiful scenery. The population was about 100,000 in 2012. It exports tea, mineral sand, agricultural products and imports grain, cement and industrial raw materials (Fig. 26).

History: it has the oldest Hindu temple in Sri Lanka. Under Dutch governance, it was one of the main ports. During World War II, it was the headquarters of Lord Mountbatten. Once an important British naval base in Asia before 1957, it is now the largest joint military, commercial and fishing port in the country.

Climate: it is a typical tropical dry and wet climate. Trincomalee usually has a dry season from March to June and a rainy season in the rest of a year, with about 1570 mm of rainfall per year. The average temperature ranges from 26 °C in December to about 30 °C in January and the hottest months of the year are from April to September.

Transportation: its road, rail and shipping transportation are very convenient. At the east end of A6 and A12 expressways and the north end of A15 expressways, Trincomalee connects with Colombo by railway with the station on the coastline near the north. Trincomalee port is the main port in the northeast of Sri Lanka and the national largest port for military, commercial and fishing.

Remote Sensing Monitoring of Trincomalee Urban Expansion from 1995 to 2015

Urban expansion characteristics of Trincomalee: it is a coastal city with urban built-up areas of 79.50 km^2, 101.14 km^2 and 119.08 km^2 in 1995, 2005 and 2015, respectively. The annual average growth rate of urban area in the two decades is 2.44% and 1.65% respectively, showing a slow growth trend (Fig. 27, Table 6 and Fig. 28).

The characteristics of land occupation during urban expansion: urban expansion was mainly concentrated in the north from 1995 to 2005 and the inland from 2005 to

Fig. 26 True color remote sensing image of Trincomalee. *Remote sensing image source* Google Earth

2015. Specifically, 13.64 and 7.76% of the total built-up land in 2005 was converted from forest land and cropland in 1995, while 5.29 and 9.81% of the built-up land in 2015 from forest land and cropland in 2005 (Figs. 29 and 30).

(a) Trincomalee remote sensing image in 1995

(b) Trincomalee remote sensing image in 2005

(c) Trincomalee remote sensing image in 2015

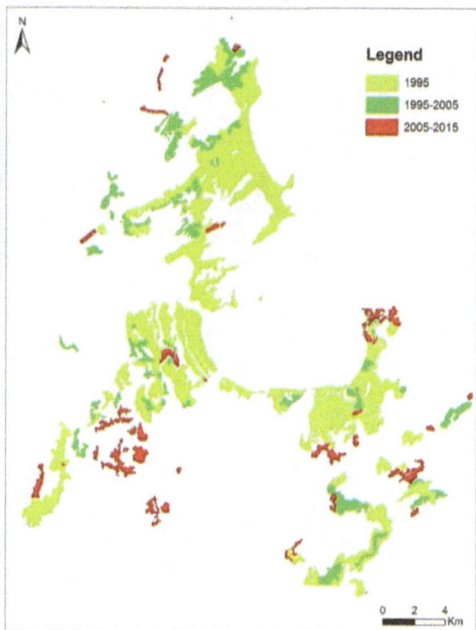

(d) built-up areas expansion of Trincomalee from 1995 to 2015

Fig. 27 Remote sensing image by years and built-up expansion image in Trincomalee. The yellow part in the Figure **d** indicates the built-up areas in 1995, the green for the urban expansion area in 1995–2005 and the red for the urban expansion area in 2005–2015

Table 6 Expansion quantity characteristics of built-up area in Trincomalee

Year	Area (km^2)	Expansion area (km^2)	Expansion speed (km^2/a)	Annual average growth rate (%)	Expansion density
1995	79.50				
2005	101.14	21.64	2.16	2.44	2.72
2015	119.13	17.99	1.80	1.65	1.78

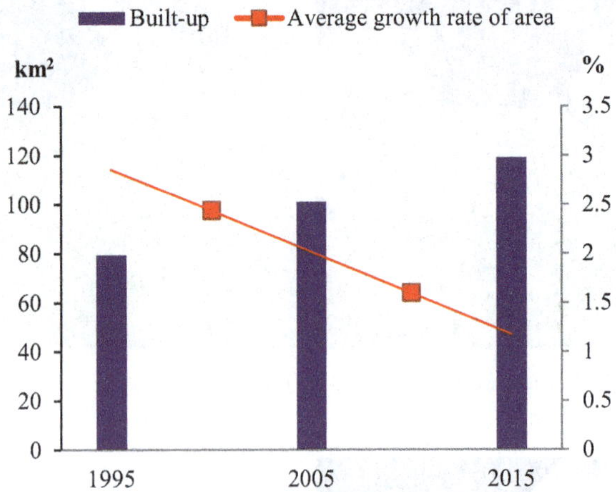

Fig. 28 Built-up area and annual growth rate of Trincomalee in 1995, 2005 and 2015

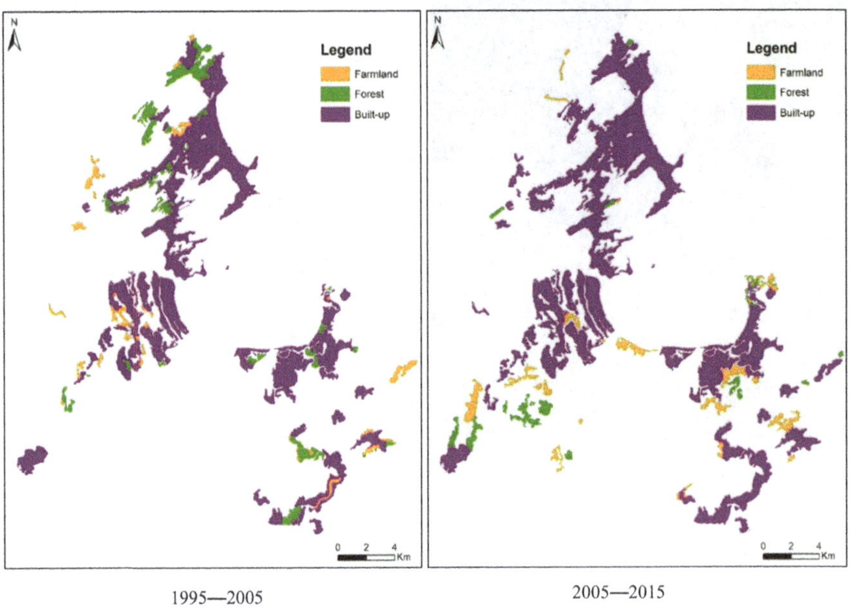

Fig. 29 Types of occupied land in urban expansion

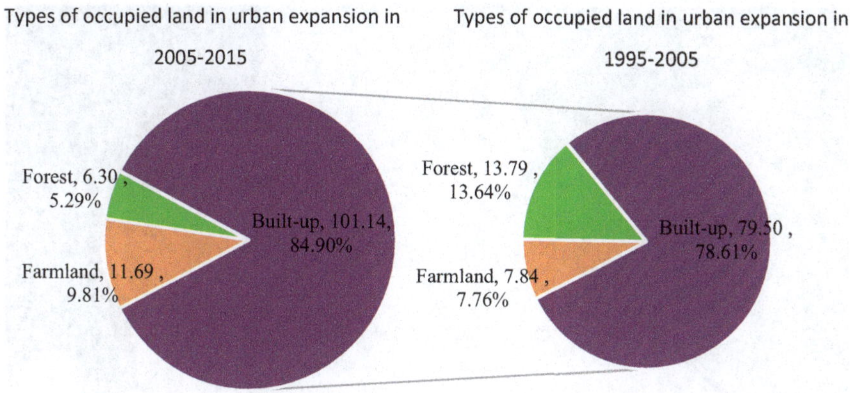

Fig. 30 Proportion of occupied land of new built-up land by periods

3.5 Anuradhapura

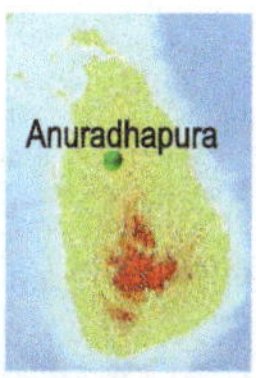

Anuradhapura, Sri Lanka's largest ancient city, is on the northern side of the Malvathu River with a population of 510,000 in 2012. 205 km north of Colombo, the city is the capital of North Central Province (Fig. 31).

History: Anuradhapura has a history of 2500 years. It became the capital of Sri Lanka in 380 B.C. and remained the royal power for 1000 years. It is the most stable and lasting political center in South Asia. The ruins of the ancient capital were discovered in the 19th century and have been repaired until now. There is a reservoir built in the 5th century for irrigation, surrounded by large rice fields and vegetable gardens.

Transportation: Anuradhapura serves as a crossroad with railways and expressways directly connecting to other main cities, such as Vavuniya, Dambulla, Puttalam, Trincomalee, Jaffna, Kurunegala and Kandy.

Culture: it is the national Buddhist center, one of the eight world heritage sites in Sri Lanka. The old district has been reserved as an archaeological garden and a holy Buddhist world, surrounded by famous pagodas and statues.

Fig. 31 True color remote sensing image of Anuradhapura. *Remote sensing image source* Google Earth

Remote Sensing Monitoring of Anuradhapura Urban Expansion from 1995 to 2015

Characteristics of urban expansion in Anuradhapura: the monitoring map and data of urban expansion changes from 1995 to 2015 show that the urban built-up areas increased by 21.91 km^2 from 1995 to 2005, with an annual average growth rate of

1.87%. Also, the spatial distribution was relatively level. However, the urban expansion slowed down significantly from 2005 to 2015 as the urban built-up areas only increased by 1.41 km^2 or 0.11% with focus on the southeast. The urban expansion in Anuradhapura is slow compared with other cities due to the local government's efforts to protect historical and cultural heritage (Fig. 32, Table 7 and Fig. 33).

The characteristics of land occupation during urban expansion: For the expansion layout, Anuradhapura mostly occupied the edging cropland and forest land from 1995 to 2005, accounting for 5.97% and 10.95% of the built-up land area in 2005,

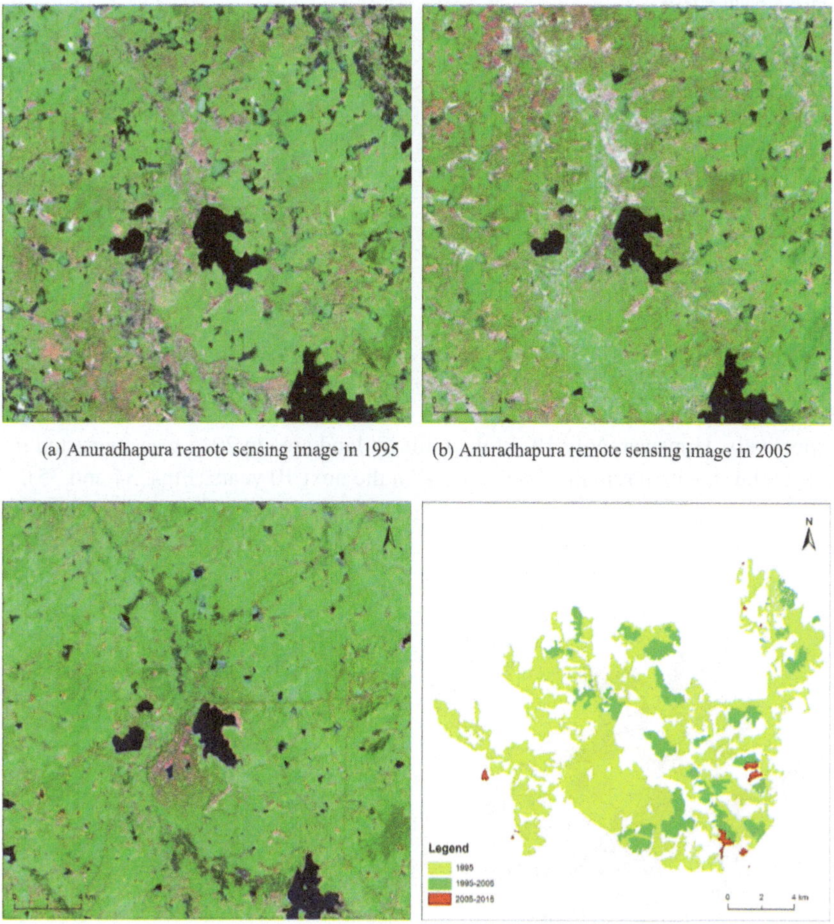

(a) Anuradhapura remote sensing image in 1995 (b) Anuradhapura remote sensing image in 2005

(c) Anuradhapura remote sensing image in 2015 (d) built-up areas expansion of Anuradhapura from 1995 to 2015

Fig. 32 Remote sensing image by years and built-up expansion image in Anuradhapura. The yellow part in the Figure **d** indicates the built-up areas in 1995, the green for the urban expansion area in 1995–2005 and the red for the urban expansion area in 2005–2015

Table 7 Expansion quantity characteristics of built-up area in compared with other cities

Year	Area (km^2)	Expansion area (km^2)	Expansion speed (km^2/a)	Annual average growth rate (%)	Expansion density
1995	107.60				
2005	129.51	21.91	2.19	1.87	2.04
2015	130.92	1.41	0.14	0.11	0.11

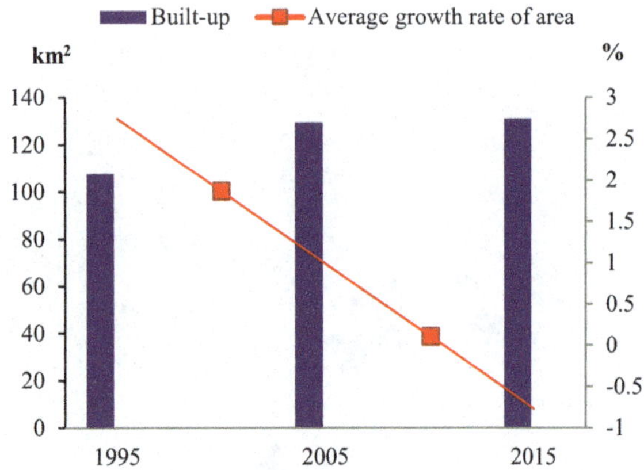

Fig. 33 Built-up area and annual growth rate of Anuradhapura in 1995, 2005 and 2015

respectively. However, only 1% of the built-up land area in 2015 was converted into built-up land, which remained unchanged in the next 10 years (Figs. 34 and 35).

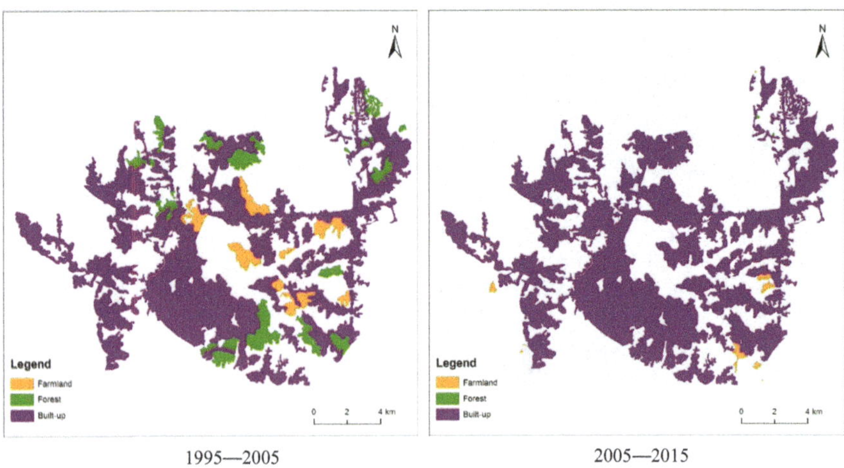

Fig. 34 Types of occupied land in urban expansion

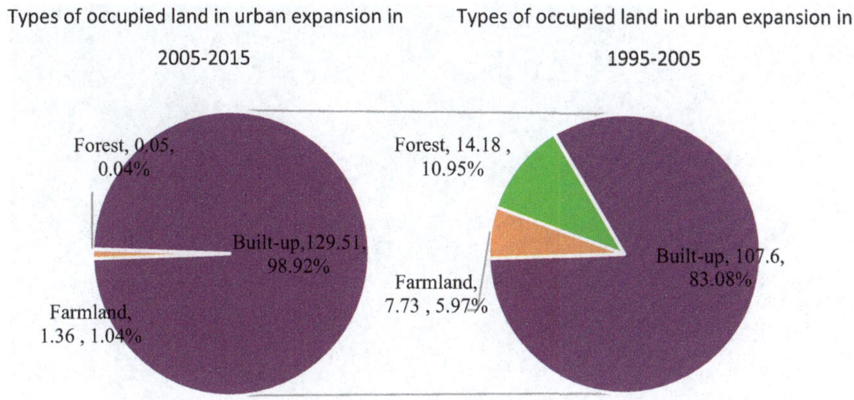

Fig. 35 Proportion of occupied land of new built-up land by periods

3.6 Polonnaruwa

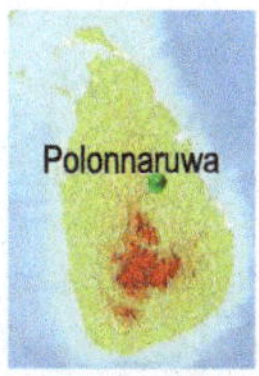

The second largest city in the North Central, Polonnaruwa is a place suitable for archeology and known as one of the cleanest cities. Its natural environment, ancient buildings and the Parakrama Samudra, a 1200-year-old artificial lake attract a large number of tourists (Fig. 36).

Location: Polonnaruwa is the main town in the North Central Province, 216 km northeast of Colombo and more than 90 km northwest of Anuradhapura.

History: Anuradhapura, the capital and Buddhist Center at that time, began to lose its power at the end of the 10th century A.D. and Polonnaruwa became the second capital until the 13th century. Later, Polonnaruwa was destroyed by the invasion of Tamils in South India and was not discovered again until the 19th century. In 1982, the ancient city of Polonnaruwa was declared a world cultural heritage.

Remote Sensing Monitoring of Polonnaruwa Urban Expansion from 1995 to 2015

Characteristics of urban expansion in Polonnaruwa: it focused on cultural protection and tourism from 1995 to 2015 with a slow urban expansion. The scale of urban built-up expansion shows the net area increased by 13.89 km^2 with a yearly growth rate of 2.81% from 1995 to 2005, while the urban expansion slowed down significantly and the rate decreased to 0.4% from 2005 to 2015 (Fig. 37, Table 8 and Fig. 38).

Fig. 36 True color remote sensing image of Polonnaruwa. *Remote sensing image source* Google Earth

Characteristics of land occupation during urban expansion: the spatial distribution of urban built-up expansion shows that the expansion of urban built-up area is distributed on the forest land in the edging area in the first decade and the area became quite small in the last 10 years. In 2005, 21.39% and 2.79% of urban built-up land was respectively converted from the forest land and cropland in 1995. In 2015, 0.89 and 3.04% of the land was from those two land in 2005. Generally, urban expansion

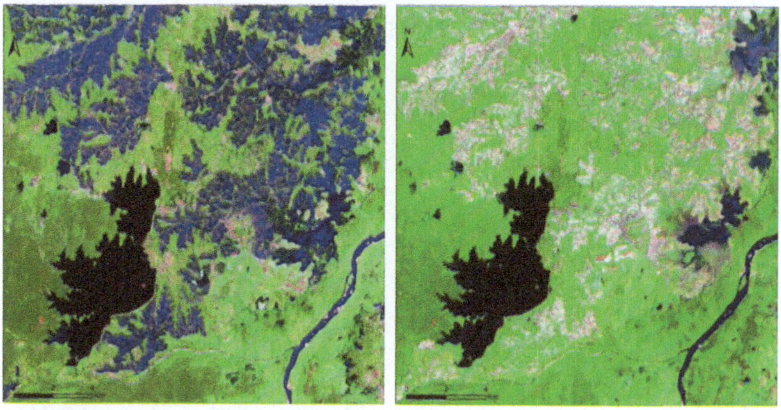

(a) Polonnaruwa remote sensing image in 1995 (b) Polonnaruwa remote sensing image in 2005

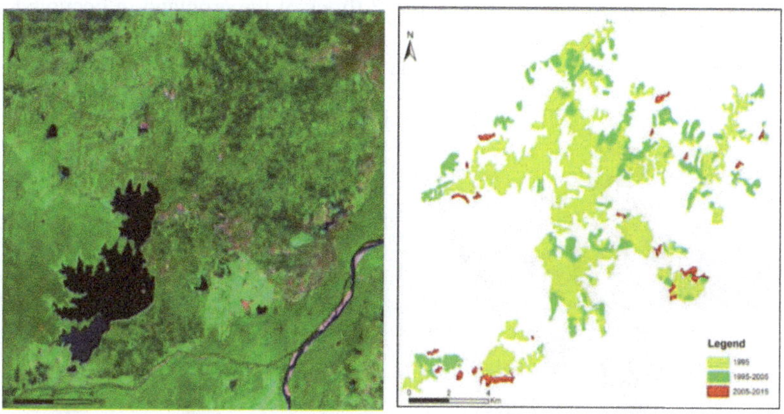

(c) Polonnaruwa remote sensing image in 2015 (d) built-up areas expansion of Polonnaruwa from 1995 to 2015

Fig. 37 Remote sensing image by years and built-up expansion image in Polonnaruwa. The yellow part in the Figure **d** indicates the built-up areas in 1995, the green for the urban expansion area in 1995–2005 and the red for the urban expansion area in 2005–2015

Table 8 Expansion quantity characteristics of built-up area in Polonnaruwa

Year	Area (km^2)	Expansion area (km^2)	Expansion speed (km^2/a)	Annual average growth rate (%)	Expansion density
1995	43.46				
2005	57.32	13.86	1.39	2.81	3.19
2015	59.66	2.34	0.23	0.40	0.41

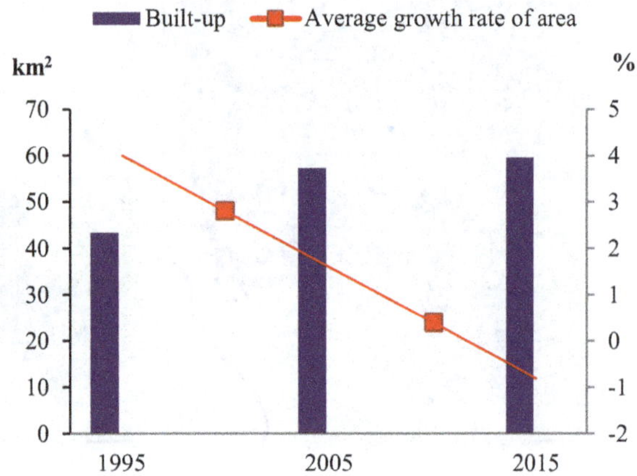

Fig. 38 Built-up area and annual growth rate of Polonnaruwa in 1995, 2005 and 2015

mainly occupied forest land, which was 3.7 times of the cropland in occupied area (Figs. 39 and 40).

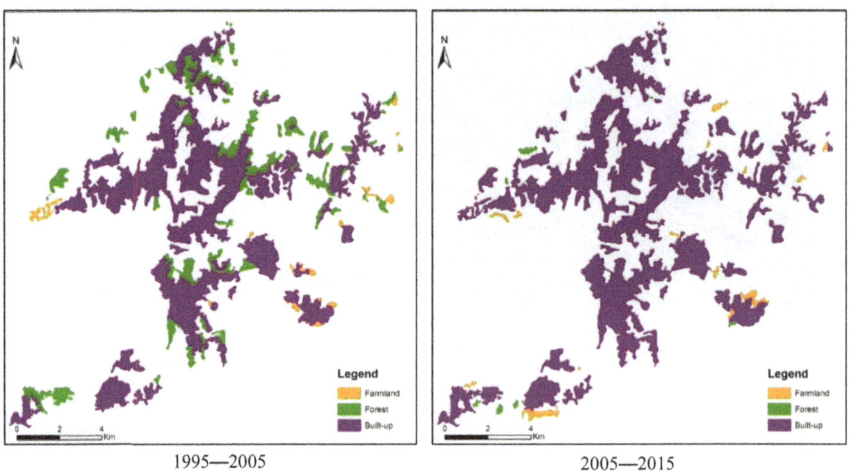

Fig. 39 Types of occupied land in urban expansion

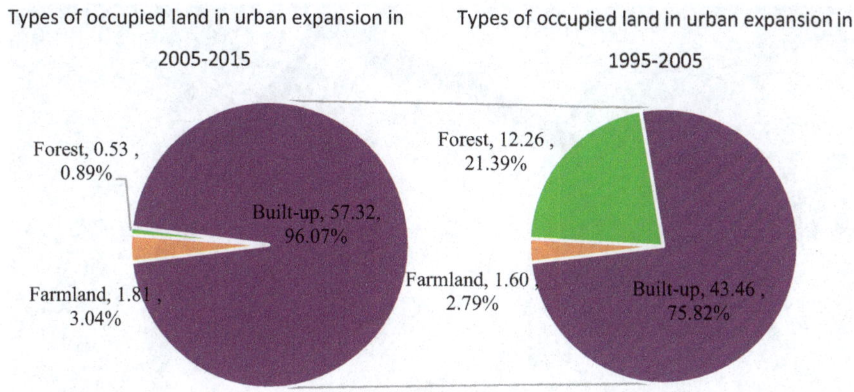

Fig. 40 Proportion of occupied land of new built-up land by periods

3.7 Kandy

Located in the Central Highlands, Kandy is an important commercial, religious, cultural and transportation center, 115 km away from Colombo with an altitude of about 500 m and a population of 125,000 in 2011. Surrounded by mountains, Kandy Lake in the center of the city is known as one of the most beautiful cities in the world (Fig. 41).

History: it was founded in the 14th century A.D., once ruled by the Sinhalese Dynasty and colonial country. Once, it was an administrative and religious center. The ancient city of Kandy, famous for its Buddhist holy land, was the last capital of the reign of King Sinhala.

Culture: Kandy retains some royal buildings and exotic colonial buildings. In the feudal era, the royal family created the essence of Sri Lanka culture, including royal subsidized temples and exquisite sculptural frescoes. The Sri Dalada Maligawa there is a famous historical building and a pilgrimage place for Buddhists.

Economy: it is the second largest city across the country and the capital of Central. Some agricultural research centers and many branches of large companies are here, such as textile, gem, furniture, information technology and jewelry industry.

Transportation: it is the main transportation hub due to its advantageous location with national roads leading to the whole country, serving as the gateway to the Central Highlands. The railway line from Colombo leads to the east coast through Kandy.

Fig. 41 True color remote sensing image of Kandy. *Remote sensing image source* Google Earth

Remote Sensing Monitoring of Kandy Urban Expansion from 1995 to 2015
Characteristics of urban expansion in Kandy: the monitoring map from shows that the expansion of urban built-up areas from 1995 to 2005 is sporadic in the urban edge while it developed inside the city from 2005 to 2015 (Fig. 42).

The urban construction is accelerating now. The urban built-up area was 33.75 km^2 in 1995, 36.69 km^2 in 2005 and 43.49 km^2 in 2015. The growth rate of urban area

(a) Kandy remote sensing image in 1995

(b) Kandy remote sensing image in 2005

(c) Kandy remote sensing image in 2015

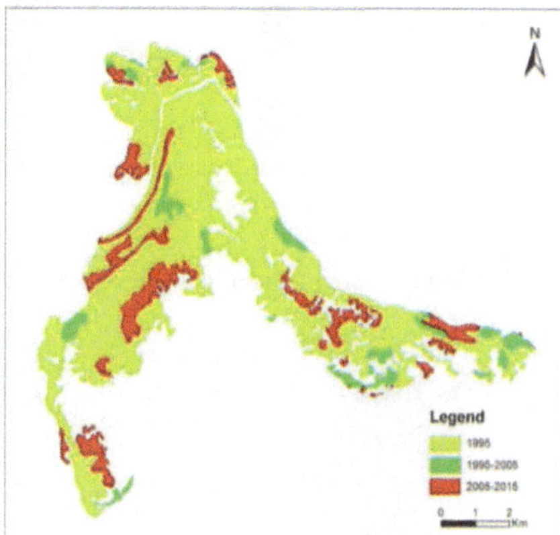

(d) built-up areas expansion of Kandy from 1995 to 2015

Fig. 42 Remote sensing image by years and built-up expansion image in Kandy. The yellow part in the Figure **d** indicates the built-up areas in 1995, the green for the urban expansion area in 1995–2005 and the red for the urban expansion area in 2005–2015

was 2.49% and 0.84% respectively. As one of the main transport hubs, it will grow more rapidly with more convenient transportation (Table 9 and Fig. 43).

The characteristics of land occupation during urban expansion: it is located in the mountainous area, thus the expansion mainly depended on the forest land. In 2005, 7.38% of the urban built-up area was converted from forest land in 1995 and only 0.64% of the area was converted from cropland. Further in 2015, 12.46% of the urban built-up area was converted from forest land in 2005 and only 3.14% of the area was converted from cropland (Figs. 44 and 45).

Table 9 Expansion quantity characteristics of built-up area in Kandy

Year	Area (km^2)	Expansion area (km^2)	Expansion speed (km^2/a)	Annual average growth rate (%)	Expansion density
1995	33.75				
2005	36.69	2.94	0.29	0.84	0.87
2015	43.47	6.78	0.68	1.71	1.85

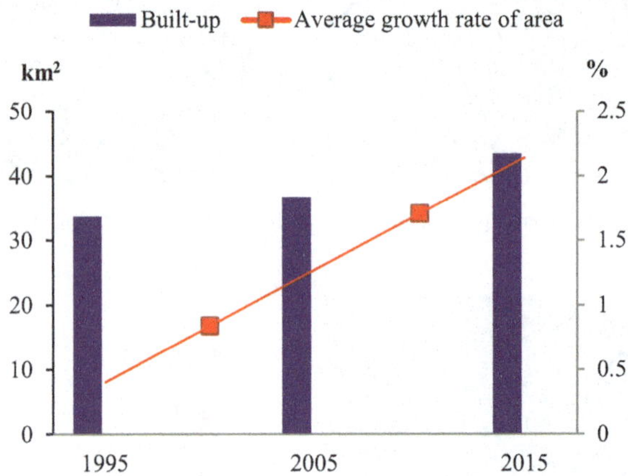

Fig. 43 Built-up area and annual growth rate of Kandy in 1995, 2005 and 2015

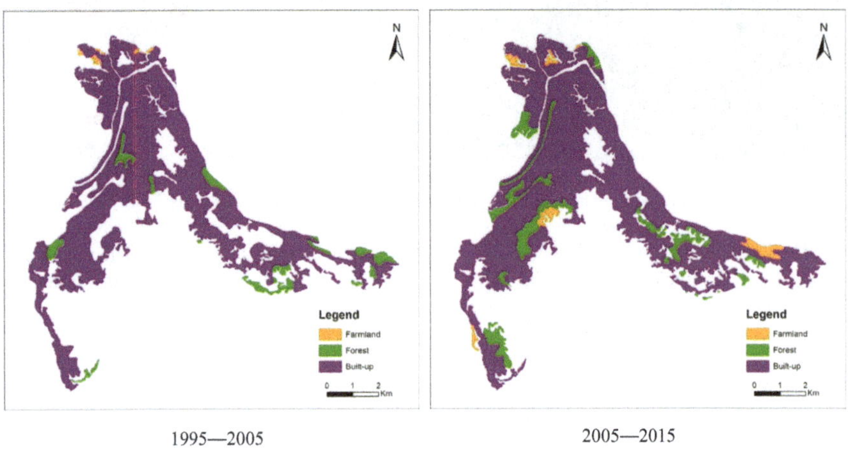

Fig. 44 Types of occupied land in urban expansion

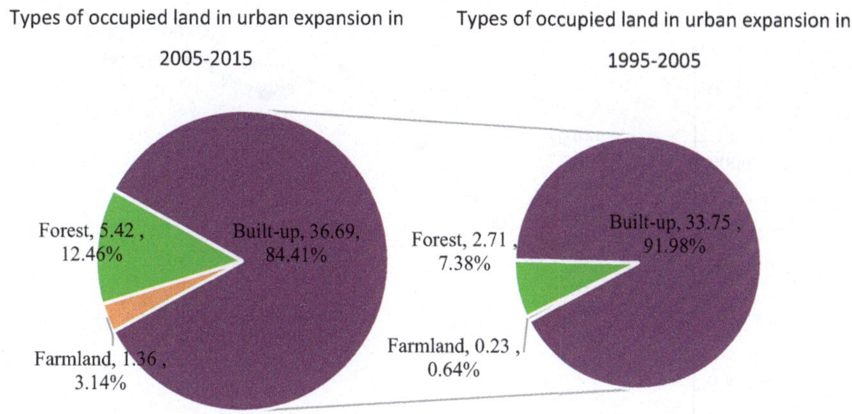

Fig. 45 Proportion of occupied land of new built-up land by periods

4 Main Crops

4.1 General Situation of Agricultural Development in Sri Lanka

Sri Lanka has a tropical monsoon climate, affected by the northeast monsoon from December to March and the southwest monsoon from June to October. The hot and humid climate and the unique local topography make it suitable for a variety of crops. The FAO data indicate that about 40% of the land is agricultural land, which was increasing from 2007 to 2014 (Fig. 46).

31.8% of the total population are farmers and the per capita agricultural land is relatively high. Consequently, personal income from agricultural production was increasing year by year. The development of agriculture has reduced the number of malnourished people and greatly improved people's livelihood (Fig. 47).

According to FAO, agricultural land has two types. The first is arable land including short-term cropland, short-term grassland for mowing or grazing, vegetable garden for market and family and temporarily idle land excluding land for leisure use. The second type is permanent cropland for long-term crops without replanting, including the land for flowering shrubs, fruit trees, nut trees and grape trees excluding the timber forests. Rice is the representative crop of arable land, while tea, fruit trees, rubber, nut trees and other flowering shrubs for profit are the representative crops of permanent cropland (Fig. 48).

It can be seen from above that in recent years, the arable land area has increased steadily, while the permanent cropland area has remained basically unchanged. Thus, the agricultural land has not changed much.

Recently, the infrastructure of large-scale plantations has gradually improved, most of which are operated by machines except for collection. Sri Lanka has a

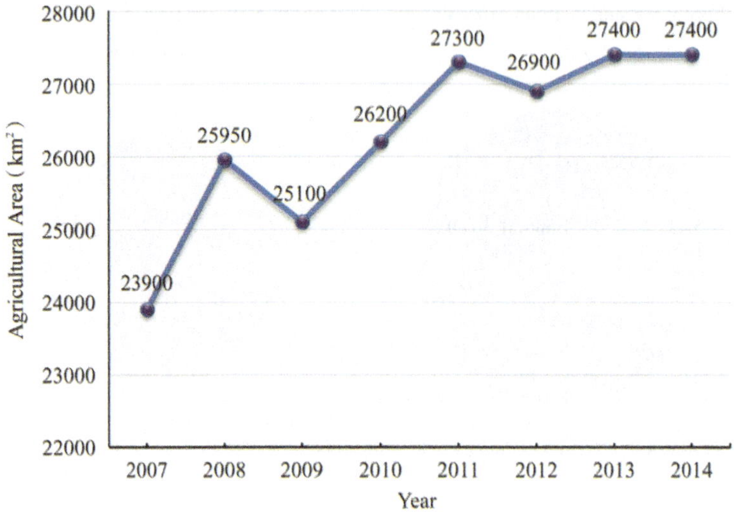

Fig. 46 Agricultural land area in 2007–2014 (km^2). *Source* World Bank Database

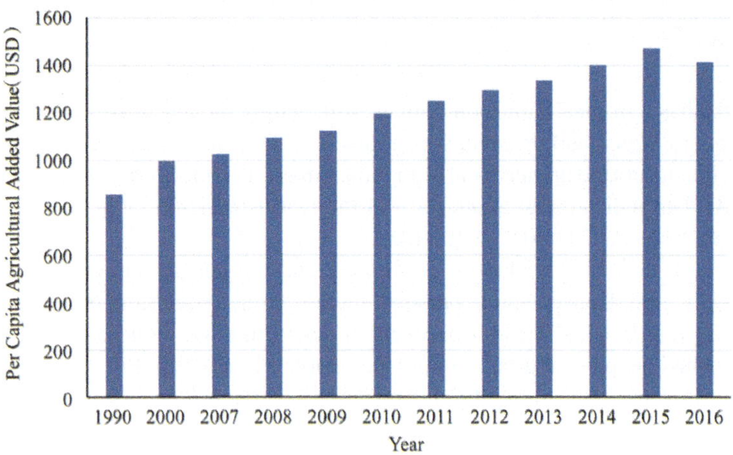

Fig. 47 Per capita agricultural-added value (USD). *Source* World Bank Database

sound transportation network for tea garden, rubber and coconut crops with railways and roads along the coast connecting Colombo, Hambantota and Trincomalee. The railway goes through the middle mountain area where tea is produced with the roads in all directions to the villages (Figs. 49 and 50).

Figure 51 the Use of Chemical Fertilizer from FAO. The data show that the chemical fertilizers for agricultural production is an important promoter of the national harvest (Fig. 52).

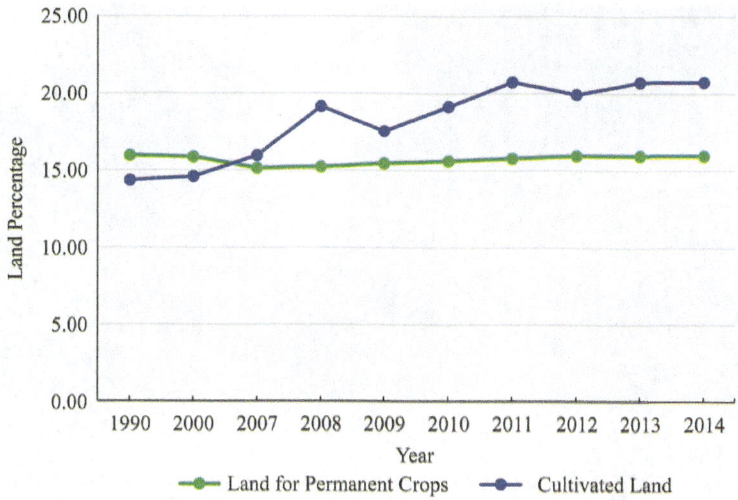

Fig. 48 Percentage of arable land and permanent cropland. *Source* World Bank Database

Fig. 49 Fields with sprinkling irrigation

The consistent agricultural classification method is issued by FAO for all countries as above. However, it can not fully describe the agricultural land and agricultural cultivation in Sri Lanka as each country has different national realities.

Therefore, the types of agricultural land will be further classified and summarized.

Fig. 50 Coconut trees planted by overlapping

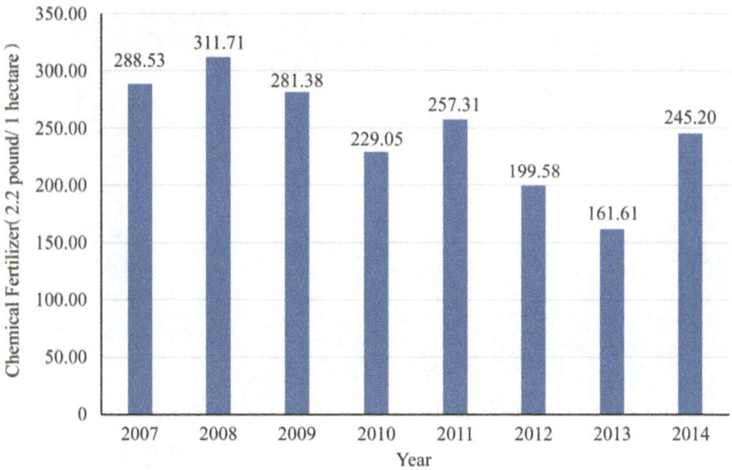

Fig. 51 Fertilizer use (kg/Ha)

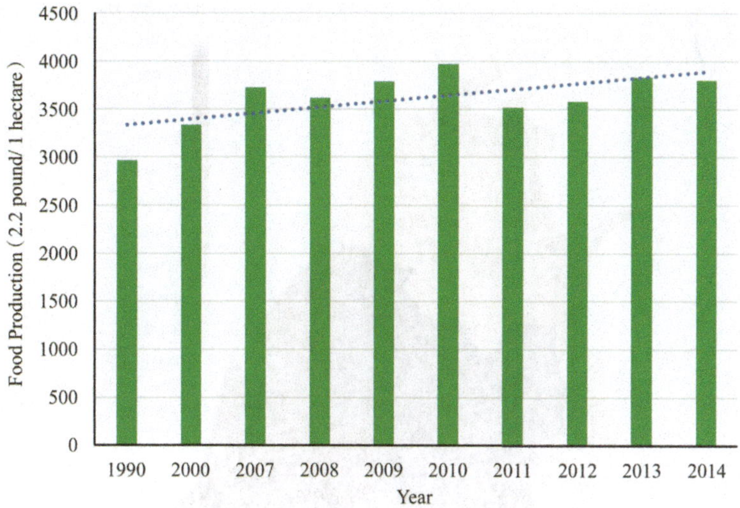

Fig. 52 Grain yield (kg/Ha)

4.2 Rice

As the main food crop in Sri Lanka, rice plays an important role in the national economy and life. It has two main cultivation seasons: Maha Season and Yala Season, referring to two monsoons. Maha Season is from September to March, while Yala Season is from May to August. Rice cannot be planted all over the country due to the uneven regional rainfall and topography.

Taking Maha Season rice planting in 2015 and 2016 as an example, the mapping of rice planting area, unit yield and total yield can directly show the general situation of rice planting (Figs. 53 and 54).

The rice is mostly planted in Anuradhapura, Kurunegala, Polonnaruwa in the middle and Ampara, Batticaloa in the east. However, the areas with the highest total yield are closely related to the direct factors such as per hectare yield and sowing area. Those are the Mahaweli River Basin and the southern Hambantota and other areas with high rainfall and rich water resources (Figs. 55 and 56).

Figure 57 the remote sensing image of rice planting area near Mahansenpura. The grid in the image with yellow brown area on the right and the green area on the left depicts rice planting fields. Due to the different growth stages of rice, the color displayed is inconsistent.

Since the national independence, rice cultivation has increased significantly, but rice production was still far lower than that of the major rice producing countries in the late 1980s. Rice planting is mainly based on small-scale farming and still adopts manual sowing, harvesting and threshing in some villages.

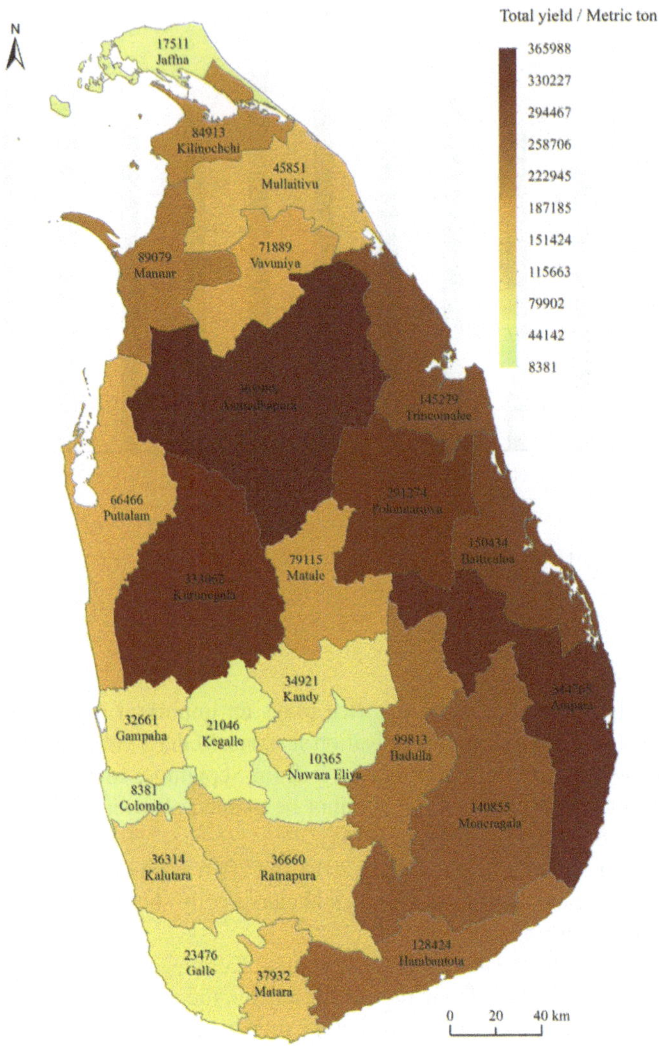

Fig. 53 Rice yield by areas. *Source* Department of Census and Statistics, Sri Lanka

In 2003, the country achieved a harvest in rice production, with the output of 38,000 tons more than the national demand. It is the first time that it has achieved self-sufficiency in rice production since independence. Sri Lanka donated 7,500 tons of rice to the World Food Programme in 2011, which means that it has transformed from a country with a shortage of rice to a country with a surplus yield (Fig. 58).

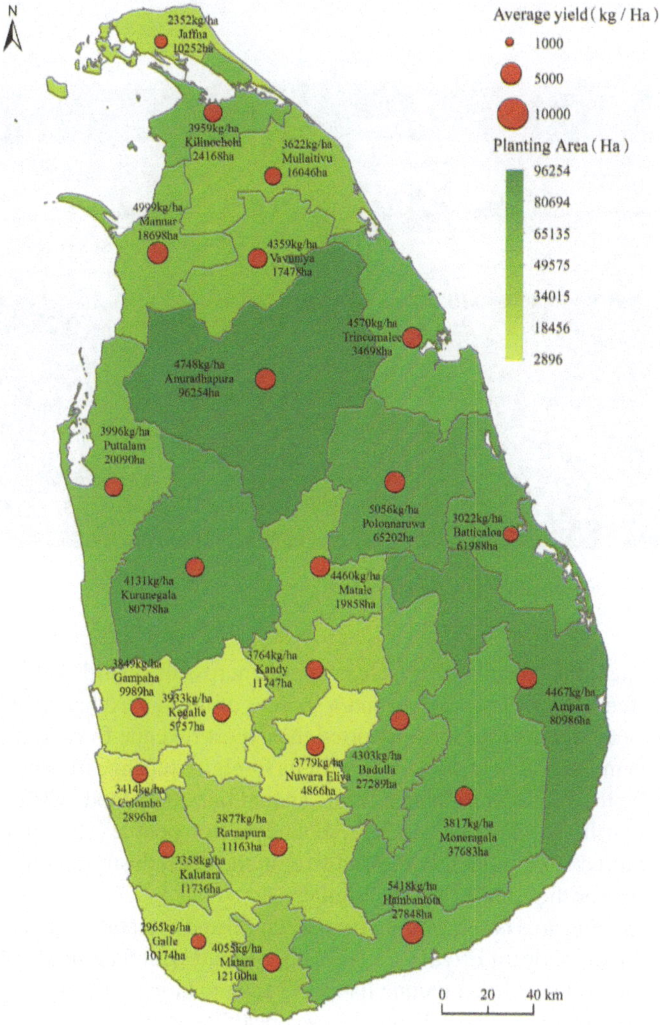

Fig. 54 Rice yield and planting area by areas. *Source* Department of Census and Statistics, Sri Lanka

4.3 Tea

Sri Lanka is the fourth largest tea producer in the world, where the Ceylon black tea is well-known. Its tea planting originated in 1867 with a history of only 150 years.

However, tea industry is its pillar industry. The tea production was 328,000 tons in 2015, of which 307,000 tons were exported, accounting for 17% of the total global

Fig. 55 Rice fields around Parakrama Samudra

tea exports then. It is the third largest tea exporter across the world, following Kenya and China. Tea industry is one of the main sources of foreign exchange in Sri Lanka. The Tea Association of Sri Lanka reported the export revenue from tea reached US $1.65 billion in 2014, accounting for 15% of its foreign exchange income. The figure surpassed the total foreign exchange for its annual import of food. The contribution rate of tea industry to GDP reached 2%. About 2 million people (10% of the national population) directly or indirectly depend live on the tea industry and thus the main way for living for thousands of local families.

The tea planting area is concentrated in the central and southern regions, especially in Nuwara Eliya. Nuwara Eliya is an elliptical plateau with an altitude of 1,902 m, where the tea produced has a unique flavor. The tea planting industry in other areas also has its own characteristics. Dimbula was one of the earliest areas growing tea in the 1870s. Tea gardens in Kandy, located between 610 and 1,219 m above sea level, had the first tea factory in the country.

There are many large tea plantations in Nuwara Eliya and small tea plantations in Galle and Ratnapura (Figs. 59, 60 and 61).

In Sri Lanka, tea is planted by the Contour Planting Method that the route of planting tea trees is in line with the contour of the land and the tea garden is usually on the hillside. Figure 62 shows the remote sensing image of Kandapola, Nuwara Eliya. The Method can be clearly displayed as the dark green area is tall trees and the light one is tea planting area.

The Central Highlands, with cool temperature and abundant rainfall, provides a suitable climate for the production of high-quality tea, serving as the main planting

Fig. 56 Remote sensing image of Parakrama Samudra and rice fields. *Remote sensing image source* Google Earth

area of tea. At the same time, some low altitude areas such as Matara and Galle have much rainfall and suitable temperature for high-quality tea.

Figure 63 shows the remote sensing image between Matara and Ratnapura. The dark green part is forest and the light one green is low-altitude tea plantation. Compared with the high-altitude area, the tea planting area in the low altitude area is more scattered.

4.4 Rubber

As early as 1876, Henerathgoda in the Gampaha, Western began to grow rubber. Latex in rubber trees can be used for rubber products and rubber wood can be utilized for furniture and other wood products. The rubber industry creates more than 200,000 jobs in rural areas including excavating latex, processing and manufacturing semi-finished rubber products and primary products. Nowadays, the country produces about 137,000 tons of rubber every year, of which the export volume accounts for 20–30% of the total output and accounts for 70–80% of domestic industrial processing.

Fig. 57 Remote sensing image of rice planting area near Mahansenpura. *Remote sensing image source* Google Earth

Fig. 58 Harvest in rice

Rubber is planted in the southwest, especially in Kegalle and Kalutara which make up more than half of the country's rubber planting area (Fig. 64).

4.5 Coconut

One of the main crops planted, coconut is the most important food security crop in Sri Lanka after rice, accounting for about 12% of all national agricultural products. According to figures released by FAO in December 2009, Sri Lanka is the fifth largest coconut producer in the world with an annual output of 2.2 million tons. Coconuts are planted all over the country with an area of 400,000 ha in 2016. About 70% of coconuts are planted for domestic consumption and about 30% for processing. Coconuts serve as the main source of edible oil and fat in daily life.

Figure 65 shows the proportion of coconut planting scale where the green sector is the proportion of small farmers and the yellow shade represent large-scale plantations. It can be seen that the coconut planting is mostly small-scale private planting and only a certain proportion of large-scale coconut planting forest land is in the central and western regions. In fact, thanks to the family estate program, residents also grow coconuts widely, especially in front of and behind their houses.

Coconut trees are tropical light crops under high temperature and rain, suitable for growth in low altitude areas, especially in coastal areas below 50 m above sea level. As a result, coconuts are less cultivated in arid areas in the north and east of

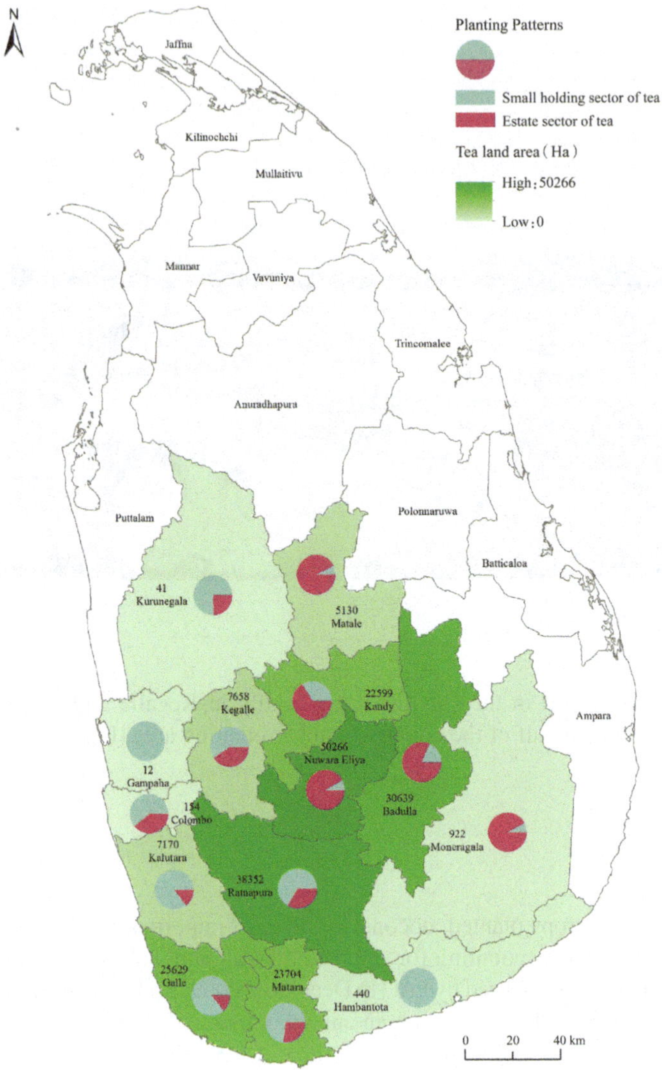

Fig. 59 Comparison of tea planting areas and different planting patterns. *Source* Department of Census and Statistics, Sri Lanka

Sri Lanka and in mountainous areas in the middle. The central and western regions are the most dominating areas of coconut planting, especially in Kurunegala, where large-scale coconut plantations and small farmers have far more planting areas than other regions (Fig. 66).

Fig. 60 The climate of central mountainous area is suitable for tea planting

Fig. 61 Tea plantations in Ella, a central mountainous city

Fig. 62 Remote sensing image of tea plantation near Kandapola, Nuwara Eliya. *Remote sensing image source* Google Earth

Figure 67 shows a large coconut forest near Mawathagama Town, Kurunegala. Compared with the natural forest, the gap between the coconut trees is large with regular distribution.

5 Tourist Attractions and World Heritage Sites

Sri Lanka is known as the Pearl of the Indian Ocean for its rich natural and cultural heritage as well as its distinctive cultural atmosphere. It houses the world's highest pagoda, the oldest bodhi tree, the largest Buddhist temple group, the relics of Buddha and charming natural scenery. The air is full of strong flavor of spices. The charm of land is based on the fragrance around the ancient temples. Moreover, the beautiful butterflies, birds and other animals make it more exotic and charming.

Sri Lanka has six world cultural heritage sites and two world natural heritage sites. Sri Lanka's richest archaeological heritage sites and the largest and most dramatic are in the Cultural Triangle. The area formed by linking the ancient capitals of Anuradhapura, Polonnaruva, and Kandy. The monuments of Sigiriya and Dambulla

Fig. 63 Remote sensing image of tea plantation between Matara and Ratnapura. *Remote sensing image source* Google Earth

are also included due to their age and artistic achievements. These three ancient capitals and two ancient sites have been listed into the World Heritage List, forming one of the world's few large heritage areas. Here was Sri Lanka's first feudal capital, the cradle of its civilization where the Sinhalese Buddhism was born. Of six world cultural heritages, 5 are located in the Cultural Triangle and the other is the Old Town of Galle.

Both world natural heritages are located in the south-central regions, Sinharaja Forest Reserve and the Central Highlands, which are rich in animal and plant resources (Table 10 and Fig. 68).

5.1 Sacred City of Anuradhapura

It is the first ancient capital and Buddhist holy land in Sri Lanka, located in the north central land and about 200 km away from Colombo.

Built in the 5th century BC with a history of 2500 years, Anuradhapura is the oldest city on the land. At that time, it was one of the domestic largest cities with an area of 1 km^2. It became the first formal capital in 380 BC and was the capital of Sinhalese Dynasty, the political and religious center of Ceylon until the 10th century AD, with enjoying a glorious history of more than 1300 years. The city was in ruins

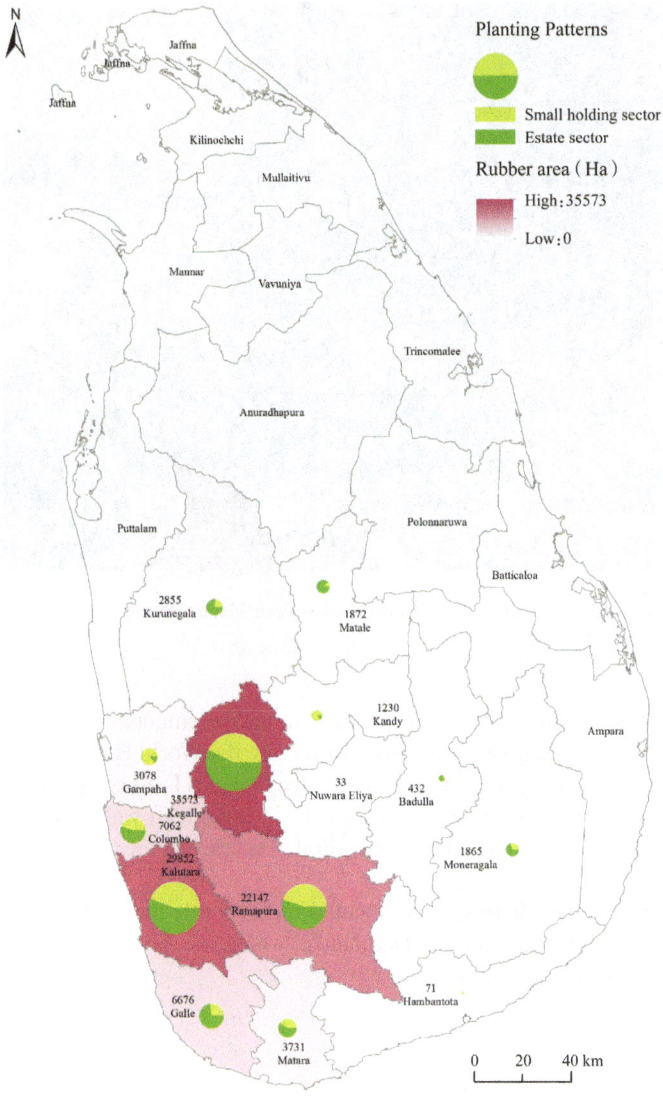

Fig. 64 Rubber planting area by areas. *Source* Department of Census and Statistics, Sri Lanka

after the invasion of South India in 993 A.D. And was discovered in the 19th century and have been repaired until now.

It serves as a holy place for Buddhist where the Buddhism prospers. More than 2000 years ago, Mahendra, the son of King Ashoka in the Maurya Dynasty, crossed the sea with Buddhist scriptures to arrive here, creating the Buddhist history in Sri Lanka. Gradually, Anuradhapura became the earliest Buddhist holy land. After thousands of years of volatility, there are still dozens of various pagodas.

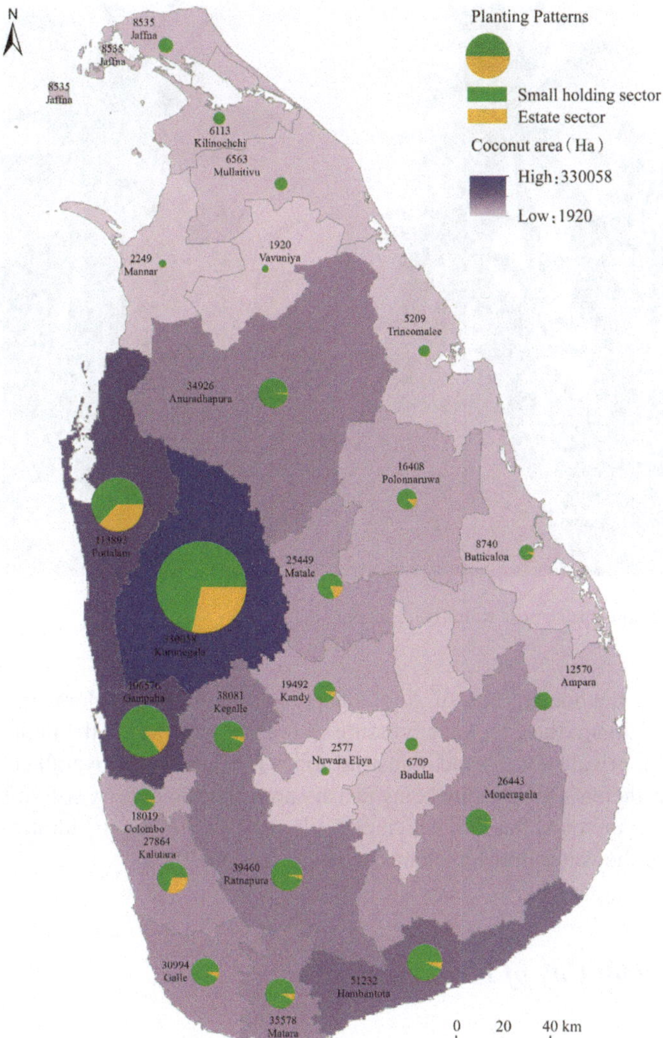

Fig. 65 Proportion of coconut planting scale by areas. *Source* Department of Census and Statistics, Sri Lanka

Anuradhapura is famous for its ancient cultural monuments. There are three main types of ruins in the ancient city, Stupa, Buddhist temple and Pokuna. The Stupa is a bell-shaped masonry building with a circumference ranging from several meters to hundreds of meters. Today, only stone platforms, foundations and pillars are left, the most famous of which is the Bronze Palace ordered by the King Dutugamunu in 164 BC. The Pokuna includes the pools for bathing and drinking, which are scattered among the trees.

Fig. 66 Coconut garden near Kalpitiya, Puttalam district

However, the ancient city of Anuradhapura faces many problems in its protection. In addition to the excessive pressure of tourists and traffic, the built and under construction private houses and industrial zones infringe on the overall architectural style of the ancient city. Despite being in a dry area, the ancient city still suffers floods frequently in the rainy season, which not only pose an erosion to cultural relics, but also pollute the water in the ancient city (Figs. 69 and 70).

5.2 Ancient City of Polonnaruwa

It is located in the northeast of Sri Lanka, 216 km away from Colombo. In 1982, UNESCO listed it as a cultural heritage in the World Heritage List.

After the destruction of Anuradhapura, the city of Polonnaruwa became the capital as well as the religious and commercial center in 933 AD. Since then, it was the second ancient capital of Sri Lanka. Polonnaruwa, the capital of the Chola and Sangala Dynasties, flourished for three centuries and was finally abandoned in the 14th century. Most of the existing monuments were built during the reign of the King Parakramabahu and Nissamka Malla. There were the Brahmanic ruins in the Kalas period and the remains of the garden city built by Pallakrama I in the 12th century.

Its architectural style and layout, quite different from that of Anuradhapura, focus on the central axis affected by the ancient capital. What is best known in the ancient city of Polonnaruwa is the Palace Complex of King Parakramabahu. To the south of

Fig. 67 Remote sensing image of coconut forest near Mawathagama, Kurunegala district. *Remote sensing image source* Google Earth

the palace was the lotus-shaped royal bath named Kumaro pool with two crocodile-shaped outlets. There was the royal conference hall on the one side of the bath, but now there are only some columns of the pavilion for visit. On the north side of the palace were many Buddhist temples and steles. Famous for its large statues, Gal Viharaya has four large Buddha statues carved from the whole stone.

Table 10 Eight world heritage list

World heritage site		Date of inscription
Culture heritage	Sacred City of Anuradhapura	1982
	Ancient City of Polonnaruwa	1982
	Ancient City of Sigiriya	1982
	Sacred City of Kandy	1988
	Old Town of Galle and its Fortifications	1988
	Rangiri Dambulla Cave Temple	1991
Nature heritage	Sinharaja Forest Reserve	1988
	Central Highlands of Sri Lanka	2010

Located near the ancient site, the Parakrama Samudra built in the 12th century is a beautiful large-scale artificial reservoir. Since ancient times, Sri Lanka has suffered from uneven distribution of rainfall in the rainy and drought seasons, which has exerted an impact on the agricultural development. After Polonnaruwa was ruled by Parakramabahu, water conservancy began to be built and several small reservoirs were merged and expanded to form a huge artificial lake. The construction of water conservancy has facilitated greatly the local area and yield for rice planting. Today, this huge reservoir still plays a role. Thus, it can be said that the Parakrama Samudra is indispensable for the prosperity of Polonnaruwa. The reservoir is surrounded by mountains with beautiful scenery (Figs. 71 and 72).

5.3 Ancient City of Sigiriya

About 170 km northeast of Colombo, it is the capital founded by Kassapa I, a patricide in 477–495 AD. of Malachite Dynasty. It is located on the Sigiriya Lion Rock, the summit of the granite mountain with a height of 180 m where people can overlook the whole jungle. The ancient city of Sigiriya was listed as a World Cultural Heritage in 1982.

In the 5th century AD, after killing his father to take the power, Prince Kassapa left Anuradhapura to Sigiriya Peak and established a palace in the new capital as he feared that his brother Maudgalya who had fled to India would come back for revenge. Eighteen years later, King Kassapa died in the battle of Maudgalya. After that, the ancient palace of Sigiriya was abandoned and was not rediscovered until resurfaced by a European archaeologist in 1894.

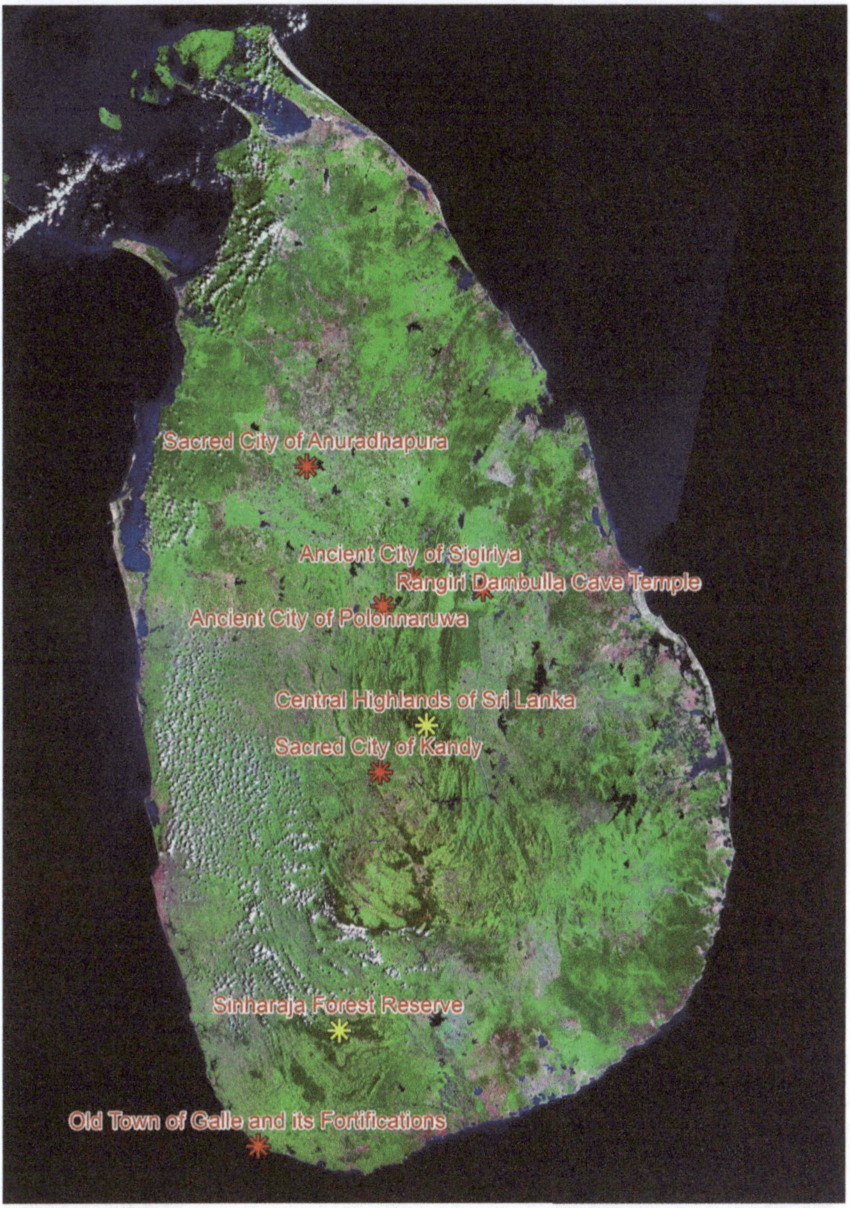

Fig. 68 Eight world heritage sites

Fig. 69 True color remote sensing image of Anuradhapura. *Remote sensing image source* Google Earth

Fig. 70 Ruvanvelisaya Dagoba. The Ruvanvelisaya Dagoba, known as the Great Dagoba, was built by King Dutugamunu, the legendary hero of Sinhala people in the 2nd century BC. It is now 55 m high and the base is 96 m in diameter. As a model of the architectural style of Sinhala, it is the most complete Buddhist monument in Anuradhapura. On the huge circular base, the five-floor Stupa is full of variety, harmony and solemnity. Known as the White Dagoba, it is one of the world's famous Buddhist relics due its pure white body and gourd-shaped gilded top

Ancient City of Sigiriya is the best-preserved city in Asia for thousands of years, where the traditional architectural style dominates. The garden and path under the shade of trees blend with the waterside pavilion, showing a perfect scenery with the symmetrical and asymmetrical elements in the varied design. A 200-m-high stone stands and two moats and three walls surround two rectangular urban areas.

The city is 3 km long and 1 km wide with great and exquisite urban planning in the 5th century AD. In line with the distant peaks, a moat runs through the entrance of the western city. The gate is made of wood and brick with flat top. The south half of the garden in the city has been excavated, while the north half with the same structure has yet to be done by archaeologists. The water park in the west city has a symmetrical layout, showing the most exquisite water pressure technology in the world since ancient times. It not only meets the needs of family horticulture and agricultural production, but also has functions in surface drainage, corrosion prevention, decoration and entertainment of waterways, protection of facilities and system refrigeration. The small water park in the west city consists of waterside pavilions, ponds, reservoirs, courtyards, fountains and waterways. Clear water flows slowly on pebbles or marbles, which is not only a practical refrigeration facility, but also a wonderful audio-visual scene. The largest water park has an island on the water from downtown, which was built five centuries earlier than that of Angkor,

Fig. 71 True color remote sensing image of Polonnaruwa. *Remote sensing image source* Google Earth

Fig. 72 Royal Palace of King Parakramabahu. The royal palace in the ancient city of Polonnaruwa is the ruins where Parakramabahu has lived, which is located on the east bank of the northern Baltic Sea. It is in the shape of rectangle, 31 m long and 12.6 m wide. It is said that it once had 7 floors high with more than 50 rooms, including audience hall, banquet hall and others supported by 36 columns of different shapes of audience hall, banquet hall and other attached halls. Nevertheless, the top floor has long disappeared as it is made of wood. Now there are only three floors of walls and a small section of steps leading to the upstairs. The wall is as thick as 3 m with regular holes. The light beam enters from the holes, which can be inferred that it should have been used to fix the floors of the second and third floors. The auditorium is famous for the eaves carved with different elephant patterns. Also, there is a beautiful lion statue at the top of the stairs

the ancient capital of Cambodia and the garden of Mughal Empire in India. On the central island stands a huge pavilion, where the water in four L-shaped ponds is connected by pipes at different depths to control different water levels. There is a large swimming pool down the wooden tunnel of the pond in the southwest. In the middle of the other small pond stands a rock, on which is a small brick Pavilion (Figs. 73, 74 and 75).

The wooden trestles in the middle of the giant rock had been damaged by weathering. Thus, one can only climb along a narrow iron ladder built recently, which is almost vertical to the cliff. Imagine that, those wooden roads must be more vulnerable and more difficult to climb than the modern ladder, as a Chinese old saying say that one man can hold out against ten thousand (Fig. 76).

Fig. 73 Remote sensing image of ancient city of Sigiriya. *Remote sensing image source* Google Earth

5.4 Rangiri Dambulla Cave Temple

As a Buddhist temple built in the 1st century BC, it was listed as a World Cultural Heritage in 1991. Situated 149 km northeast of Colombo and 60 km north of Kandy, it is located in the middle of Sri Lanka close to the ancient city of Sigiriya. It is regarded as the pilgrimage place in Sri Lanka. The typical Cave Temple on the slope of a stone mountain was carved from the rock, where there are all kinds of stone Buddha statues and other sculptures as well as murals around. With five holy halls, it is the largest and most preserved cave temples in Sri Lanka. The Temple is as famous as Indonesia Borobudur, Cambodia Angkor and India Ajanta Grottoes, which are also world natural and cultural heritage, which are all of great religious and artistic value and show the charm of Buddhist art.

The Temple was considered as a cemetery 2700 years ago. However, the caves flourished after Buddhism was introduced. As a temple, it can be dated back to the 1st century BC. It is generally believed that King Valagamba of Anuradhapura transformed these caves into temples at that time. In order to avoid the assassination from South India, he exiled here and hid for 15 years. After returning to power, he built the temple in Dambulla to show his gratitude. Later, many kings followed his step to build temples here. By the 11th century, it had become one of Sri Lanka's major Buddhist centers and it remains the name now. In 1190, Nissanka Malla of the

Fig. 74 Remote sensing image of Sigiriya Lion Rock. Sigiriya Lion Rock, located in the center of the ancient city of Sigiriya, is a palace built on the orange giant rock. The huge rock is about 200 m high on the enormous artificial lake. At the foot of the rock lies a plain. At the top, a palace is built on the platform of 1.2 ha, where there are halls, rooms, platforms, ponds and gardens full of flowers. In ancient times, the way to the palace is only stairs and corridors built on the cliffs. The remote sensing image can show a dense jungle around Sigiriya Lion Rock and even a delicate planning of the castle in the air on the top of the rock with a high resolution. *Remote sensing image source* Google Earth

Fig. 75 Lion's paw gate

Fig. 76 Ruins of palace complex on the top of Sigiriya Lion Rock

Polonnaruwa Dynasty gilded the cave and built 70 new Buddha statues. During the Conti Dynasty in the 18th century, it was restored.

As a famous tourist attraction and religious place, the Dambulla Cave has attracted many tourists and pilgrims that bring pressure to the protection. The excessive visitors make it difficult to protect and manage the cultural relics and the flash photography may also cause damage to the color of Buddha statues and murals. Today, the protection of the Temple focuses on the murals, involving the cleaning and protective coating of murals. But while protecting cultural relics, we should also pay attention to the inheritance of painting techniques using traditional natural dyes, so as to facilitate the future conservation and restoration of cultural relics (Figs. 77 and 78). (Wikipedia)

Fig. 77 True color remote sensing image of Rangiri Dambulla Cave Temple. The temple is located on a huge rock, more than 160 m above the surrounding ground. The bottom area of the rock is 1.5 km^2 and the top area is 550 m^2. There are more than 80 caves. The majority are distributed in five separate caves, each of which has a large number of Buddha statues and other statues. Besides, the roof of the temple and the surrounding walls are decorated with numerous colorful murals, mostly about Buddhist stories. Totally, there are 2,100 m^2 of murals and 160 statues, extremely precious. The statues include 153 Buddhas, 3 Kings and 4 deities. *Remote sensing image source* Google Earth

Fig. 78 Golden Temple. Built in 2000, the modern Golden Temple is located at the piedmont of Rangiri Dambulla Cave Temple, with a 30-m-high golden Buddha statue on the top

5.5 *Sacred City of Kandy*

The capital at the end of Sinhala Dynasty, it is the third and the last capital of Sri Lanka in the era of ancient king. It was listed as a World Cultural Heritage in 1988.

Kandy is famous for its Buddhist culture where the Sri Dalada Maligawa is a famous Buddhist pilgrimage land with the Buddha's tooth. It was founded in the 14th century and invaded by the British in 1815. Lasting for five centuries, it accumulated human wisdom in urban planning, architecture, sculpture, painting, music and dance, so it was considered as an important cultural heritage.

The city is located in a valley 488 m above sea level of the central mountainous area, 115 km northeast of Colombo. To the south, it is adjacent to the artificial Kandy Lake. Around the city, Mahaweli River is the longest river in Sri Lanka. The city is not only beautiful, but also of great strategic significance. It is a castle easy to defend and hard to attack. The palaces of the last king are located on a hillside terrace in the city, with dense trees on the hillside at the back and defense walls and trenches at the front. With atrium as the axis, the palace is divided into courtyards, including the King's palace, the Queen's living room, the King's gun house and the Queen's bathroom.

The most important part there is Sri Dalada Maligawa on the bank of Kandy Lake, which is close to the King's palace for storing the Buddha Sakyamuni's tooth relics. Built in the 15th century, it experienced renovation and expansion consecutively by

Kings, so it became magnificent. Built on a platform about 6 m high, the temple has a main entrance at the west gate surrounded by the temple river. It is divided into two layers, with a complex hall structure including the Buddha hall, drum hall, long hall, chanting hall, treasure house, inner hall, etc. The center is the inner hall on the second floor, which is dedicated to the national treasure, Buddha teeth (Figs. 79 and 80).

Fig. 79 Remote sensing image of sacred city of Kandy. *Remote sensing image source* Google Earth

Fig. 80 Sri Dalada Maligawa

5.6 Old Town of Galle and Its Fortifications

Located in the southwest of Sri Lanka, the old town of Galle is on the coast of the Indian Ocean about 100 km south of Colombo. Old Town of Galle and its Fortifications was listed as a World Cultural Heritage in 1988.

Built in the 16th century by the Portuguese, it reached the peak in the 18th century but began to decline after the British invasion. Constructed by southern Europeans and Southeast Asians, it is a typical fortress city integrating the traditional architectural styles of Europe and South Asia.

Old Town of Galle and its Fortifications are on the rocky peninsula, which is a natural port surrounded by the sea on three sides. Its geographical advantages make Galle become one of the important hubs from Europe to the Far East in the era of great navigation. The peninsula extends from south to north. The inner layout of the fortifications designed by the Dutch in 1669 is in a straight grid with a wall of 3 km long. In order to prevent the Portuguese from retaking Galle, the Dutch built a wall of up to six or seven meter along the three sides of the coast. In 1873 in order to improve the transportation, the British colonists at that time added a door on the north wall for the carriage to pass through. The city wall and its gate have been listed as World Cultural Heritage by UNESCO. UNESCO described it as a whole city that describes the combination of European architecture and South Asian tradition in the 16th–19th century.

The main institutions and facilities are the governor's residence, the cabinet office, the town hall, the court, the residence of the administrative officer, the tax office, the exchange, the ammunition store, the blacksmith shop, the labor hospital, the ship workshop, the navy guard barracks, the coconut warehouse, the rope yard, the mosque, the Javanese residence as well as various warehouses and cellars. Most buildings face the street and have back gardens or backyards with narrow entrance.

Galle is now growing rapidly and needs a more balanced protection of the ancient town and the city. Experts from the International Committee on Monuments and Sites also suggested that the world heritage and its buffer zone should be clearly defined, the standards for the protection should be increased and scientific research and architectural details should be carried out for the development of the historic city. Furthermore, sufficient technical personnel should be recruited to draw up and implement the overall development plan, so that the development of the historic city can be guided and the international protection regulations can be followed for the proper protection and restoration. In addition, the development of Galle port requires more in the protection of local cultural heritage. For example, how to balance the future construction of Galle port, traffic development, planning and arrangement of port logistics and how to better protect of Galle Port.

Since Galle is located on the edge of the Indian Ocean, its protection also needs to consider the impact of the ocean. In addition to the surface archaeology and protection, the underwater archaeology should also be paid attention to (Fig. 81).

5.7 Sinharaja Forest Reserve

Established in 1978 and listed as World Natural Heritage in 1988, Sinharaja Forest Reserve is one of the national parks.

It is located in the border area between Sabaragamuwa and Southern in the southwest, 21 km from the east to the west, 7 km from the north to the south, 300 to 1170 m above sea level. It has a core protection area of 85.64 km^2. Sinharaja Forest Reserve is the only existing primitive tropical rainforest in Sri Lanka. The vegetation includes tropical lowland rain forest and tropical moist evergreen broad-leaved forest. More than 60% of the trees here are endemic and rare. It houses many local wild animals, especially birds as it is the home of 95% of Sri Lanka's endemic birds. The reserve is also home to more than 50% of Sri Lanka's mammals and butterflies as well as various insects, reptiles and rare amphibians.

Currently, the threat to the value and integrity of the reserve comes from farmland encroachment, especially along the southern boundary of the reserve. In addition, the development indirectly promoted the extension of the road and then opened up the routes and access to the reserve. Nevertheless, illegal logging and resource transfer are also born, posing a threat to the reserve. 13 other adjacent natural forest areas surrounding the Sinharaja Forest Reserve provide an additional protection for

Fig. 81 Remote sensing image of Galle town. *Remote sensing image source* Google Earth

Fig. 82 Remote sensing image of Sinharaja forest

its boundary, which however needs further definition and definition. The management organization is also trying to further improve the protection of the reserve by managing the surrounding areas, so as to further reduce the impact of intensive land use on the reserve (Fig. 82). (UNESCO)

5.8 Central Highlands of Sri Lanka

The Central Highlands are located in the south-central region with the tropical marine climate. The mountainous terrain, sufficient sunshine and abundant rainfall make the local vegetation flourish, which breeds a very rich animal and plant resources.

It was listed as a world heritage site in 2010. The site includes Peak Wilderness Protected Area, Horton Plains National Park and Knuckles Conservation Forest. These mountain forests at an altitude of 2500 m have abundant animal and plant resources, including endangered species such as western-purple-faced langur, Horton Plain slender loris and Sri Lankan leopard. The region is considered to be a hotspot of biodiversity (Fig. 83).

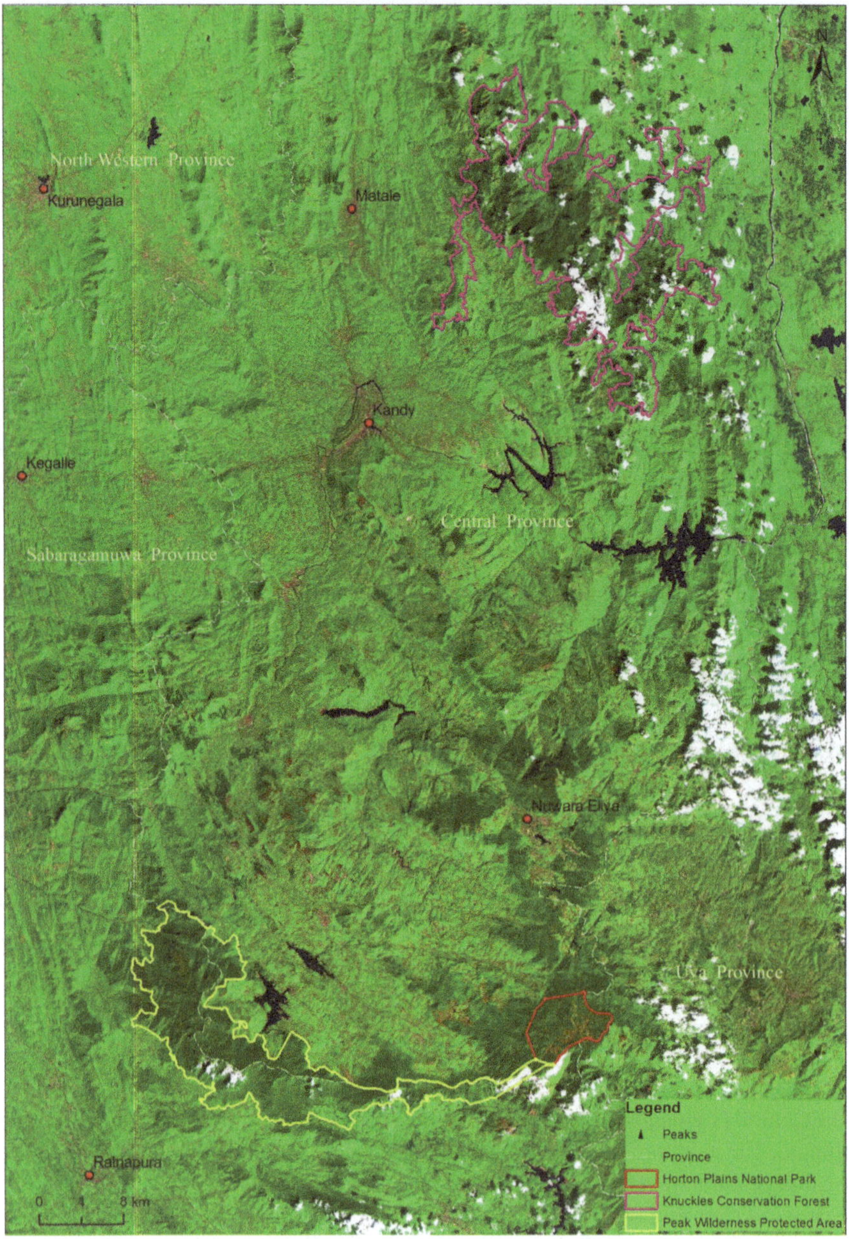

Fig. 83 Remote sensing image of the Central Highlands

Peak Wilderness Protected Area is a tropical rainforest surrounded by Sri Pada Mountain, Adam peak, covering an area of 224 km^2. The altitude of the reserve is 300–2240 m with large fluctuations. It has distinctive geographical characteristics with many high mountains and high waterfalls as well as the birthplace of many rivers. Adam Peak, or Sri Pada as known in Sri Lanka, is 2243 m above sea level. The mountain features a rock formation that is considered as the footprint of Buddha for Buddhist, Shiva for Hindus and Adam for Muslims and Christians. It is a famous Buddhist mountain where there are three channels for tourists and believers in the reserve. Except for that, all areas are controlled by Sri Lanka Wildlife Protection Agency without any tourism facilities to maintain the purity of the forest.

Horton Plains National Park is an important eco-tourist place covered by mountain grassland and cloud forest. Located in the south end of the Central Highlands and the east side of Peak Wilderness Protected Area, it is 32 km away from Nuwara Eliya in the north with an area of 3160 ha. The Horton is actually a plateau with an altitude of 2100–2300 m. The second peak of Sri Lanka, Kirigalpotta (2395 m above sea level) and the third peak, Totapola (2359 m above sea level), are located in the west and north respectively. It is the birthplace of Mahaweli River, Kelanni River and Walawe River and the most important watershed in Sri Lanka. The vegetation there is mainly grassland as well as mountain forest and woody plants. The animal is composed of mammals such as sambar deer, and long clawed shrew, purple faced langur, Sri Lanka leopard, etc. The red slender loris is the unique species here and one of the most endangered species in the world. It is also an important bird reserve with 21 species of birds that can only be seen in Sri Lanka living in Houghton plain and its nearby Wilderness Reserve. The World's End and Baker's falls are the most famous tourist attractions here. The former one is a cliff with a vertical drop of 870 m on the south boundary of the park (Figs. 84, 85 and 86).

The Knuckles Conservation Forest is the epitome of Sri Lanka's climate with all the climate types. There is an isolated cloud forest at high altitude with a large number of animal and plant species, some of which are unique in the world. Although the area is only 0.03% of the country's number, it features the high biodiversity. More than 34% of Sri Lanka's unique trees, shrubs and herbs can be found here.

Fig. 84 Horton Plains National Park

Fig. 85 World's end

Fig. 86 Three-dimensional remote sensing image of world's end. *Remote sensing image source* Google Earth

The manufacturer's authorised representative in the EU is Springer Nature Customer Service Centre GmbH, Europaplatz 3, 69115 Heidelberg, Germany. If you have any concerns regarding our products, please contact ProductSafety@springernature.com

Printed and bound by CPI Group (UK) Ltd, Croydon, CR0 4YY
30/09/2026
02225297-0012